T0295332

Advances in Manufacturing Technology

Sustainable Manufacturing Technologies: Additive, Subtractive, and Hybrid
Series Editors: Chander Prakash, Sunpreet Singh, Seeram Ramakrishna, and Linda Yongling Wu

This book series offers the reader comprehensive insights of recent research breakthroughs in additive, subtractive, and hybrid technologies while emphasizing their sustainability aspects. Sustainability has become an integral part of all manufacturing enterprises to provide various techno-social pathways toward developing environmental friendly manufacturing practices. It has also been found that numerous manufacturing firms are still reluctant to upgrade their conventional practices to sophisticated sustainable approaches. Therefore this new book series is aimed to provide a globalized platform to share innovative manufacturing mythologies and technologies. The books will encourage the eminent issues of the conventional and non-conventual manufacturing technologies and cover recent innovations.

Advances in Manufacturing Technology
Computational Materials Processing and Characterization
Edited by Rupinder Singh, Sukhdeep Singh Dhami, and B. S. Pabla

For more information on this series, please visit: https://www.routledge.com/
Sustainable-Manufacturing-Technologies-Additive-Subtractive-and-Hybrid/
book-series/CRCSMTASH

Advances in Manufacturing Technology

Computational Materials Processing and Characterization

Edited by

*Rupinder Singh, Sukhdeep Singh Dhami,
and B.S. Pabla*

CRC Press
Taylor & Francis Group
Boca Raton London New York

CRC Press is an imprint of the
Taylor & Francis Group, an **informa** business

First edition published 2022
by CRC Press
6000 Broken Sound Parkway NW, Suite 300, Boca Raton, FL 33487-2742
and by CRC Press
2 Park Square, Milton Park, Abingdon, Oxon, OX14 4RN

Library of Congress Cataloging-in-Publication Data

Names: Singh, Rupinder, editor.
Title: Advances in manufacturing technology : computational materials processing and
characterization / edited by Rupinder Singh, Sukhdeep Singh Dhami, and B. S. Pabla.
Description: First edition. I Boca Raton, FL : CRC Press, [2022] I
Series: Sustainable manufacturing technologies: additive, subtractive, and
hybrid series I Includes bibliographical references and index.
Identifiers: LCCN 2021046487 (print) I LCCN 2021046488 (ebook) I
ISBN 9781032067476 (hbk) I ISBN 9781032067490 (pbk) I ISBN 9781003203681 (ebk)
Subjects: LCSH: Manufacturing processes.
Classification: LCC TS183 .A394 2022 (print) I LCC TS183 (ebook) I
DDC 670--dc23/eng/20211012
LC record available at https://lccn.loc.gov/2021046487
LC ebook record available at https://lccn.loc.gov/2021046488

ISBN: 978-1-032-06747-6 (hbk)
ISBN: 978-1-032-06749-0 (pbk)
ISBN: 978-1-003-20368-1 (ebk)

DOI: 10.1201/9781003203681

Typeset in Times LT Std
by KnowledgeWorks Global Ltd.

Contents

SECTION I   Computational Material Processing

SECTION II   Material Characterization

SECTION III   Smart Manufacturing

Preface

In the past two decades, there has been a major change in manufacturing technology in terms of material processing, ergonomic-based designs, finite-element analysis, automation, and use of the Internet of Things. Also, nanostructure materials/ nanocomposites have become increasingly important because of growing areas of applications in smart manufacturing. Furthermore, there is an increasing significance of robotics, automation, and 3D printing processes in light of Industry 4.0. This book is a cross-disciplinary summit on advances in manufacturing technology from a computational material processing viewpoint. In other words, this project is focused on the design of experiment-based computational models. These models involve finite-element analysis along with ergonomic-based design of tooling, for both conventional and unconventional manufacturing processes. Regarding smart manufacturing, this book outlines solutions for 3D-printed thermoplastic parts.

A few figures have been purposefully repeated to avoid a break in thoughts during reading to search for them elsewhere in the book. It is expected that the book will be useful for not only the Bachelor's/Master's degree-seeking students and research scholars, but also for the practicing engineers in this field.

In spite of our best efforts, there are bound to be some mistakes. The same may kindly be brought to our attention for rectification. Any other suggestions to improve the book will be thankfully acknowledged.

I sincerely wish that the book may meet the expectations of the readers on this vital subject of manufacturing technology.

Rupinder Singh, S.S. Dhami, and B.S. Pabla

About the Authors

Rupinder Singh received his PhD degree in mechanical engineering in 2006 from T.I.E.T Patiala, Masters of Technology (Production Engineering) in 2001, and Bachelors of Technology (Production Engineering) in 1999 from Punjab Tech. University. He is a Chartered Engineer by the Institution of Engineers (India) and has been awarded the AICTE Young Teacher award, DST Young Scientist award, and UGC Research award. Currently, he is a professor at the Department of Mechanical Engineering, National Institute of Technical Teachers Training and Research, Chandigarh, India. He has more than 20 years of teaching and research experience in production and industrial engineering, with a special emphasis in additive manufacturing, nonconventional machining, and casting. He has guided 22 PhD and 102 Master's students, as well as coordinated 19 financed research projects. He has received several scientific awards. Professor Singh has worked as an evaluator of projects for international research agencies as well as an examiner of PhD theses for many Indian universities. He is the guest editor for several international journals, a book editor, and advisor for many international conferences. In addition, as author and co-author, he has published more than 20 monographs, 85 book chapters, and 550 articles in journals and conferences (more than 300 articles in journals indexed in SCI (SCOPUS h-index:36+/5387+ citations).

Sukhdeep Singh Dhami works as a Professor in the mechanical engineering department in the National Institute of Technical Teachers Training, Chandigarh, India. He has more than 25 years of experience in the areas of automation, mechatronics, and applications of AI-based tools in the domain of mechanical engineering. Professor Dhami has supervised numerous PhD and Master's level candidates in the field of manufacturing, automation, and condition monitoring. He has published numerous research papers in reputed journals and presented his research work in international conferences in the United States, Australia, Japan, and China. He is also coordinating funded research projects.

B.S. Pabla works as a Professor in the mechanical engineering department in the National Institute of Technical Teachers Training, Chandigarh, India. He has more than 35 years of experience in industry, teaching, and research. His areas of interest are manufacturing technology, computer integrated manufacturing systems, and engineering optimization. He is also the principal investigator for a large number of sponsored research projects. He has guided a number of PhD and Master's level students and published research papers in reputed international journals and conferences.

Contributors

Mohammed Ali
Aligarh Muslim University
Aligarh, India

Navneet Singh Bhangu
Guru Nanak Dev Engineering College
Ludhiana, India

Rahul Bhardwaj
CSIR-CSIO
Chandigarh, India

Chandrakant Chaturvedi
Department of Mechanical Engineering
National Institute of Technical Teachers
    Training and Research
Chandigarh, India

Pratyay Choudhury
U.I.E.T., Panjab University
Chandigarh, India

Nisha Chugh
Panipat Institute of Engineering and
    Technology
Panipat, India

Soumyajit Das
School of Mechanical Engineering
KIIT (Deemed to be University)
Bhubaneswar, India

Mandeep Kaur Dhami
Sri Guru Gobind Singh College
Chandigarh, India

Sukhdeep S. Dhami
National Institute of Technical Teachers
    Training and Research
Chandigarh, India

Amit Kumar Dubey
Panipat Institute of Engineering and
    Technology
Panipat, India

Abdul Faheem
University Polytechnic
Aligarh Muslim University
Aligarh, India

Sandip Singh Gill
National Institute of Technical Teachers
    Training and Research
Chandigarh, India

Sandeep Grover
JC Bose University of Science and
    Technology
YMCA
Faridabad, India

Abhijeet Gupta
Hi-Tech Institute of Technology
Khurda, Odisha, India

Anjali Gupta
U.I.E.T., Panjab University
Chandigarh, India

Faisal Hasan
ZHCET, Aligarh Muslim University
Aligarh, India

Ravreet Kaur
U. I. E. T, Panjab University
Chandigarh, India

Mohd. Gulam Waris Khan
Mechanical Engineering Section
University Polytechnic, Aligarh Muslim
    University
Aligarh, India

Mohd. Yunus Khan
National Institute of Technical Teachers
    Training and Research
Chandigarh, India

Banoth Krishna
National Institute of Technical Teachers
    Training and Research
Chandigarh, India

Harmesh Kumar
U.I.E.T., Panjab University
Chandigarh, India

Parveen Kumar
National Institute of Technical Teachers
    Training and Research
Chandigarh, India

Prashant Kumar
U.I.E.T., Panjab University
Chandigarh, India

Rajesh Kumar
U.I.E.T., Panjab University
Chandigarh, India

Vikas Kumar
Department of Mechanical Engineering
U.I.E.T., Panjab University
Chandigarh, India

Vinod Kumar
Punjabi University
Patiala, Punjab

Mansi
National Institute of Technical Teachers
    Training and Research
Chandigarh, India

Vijay Kumar Meena
CSIR-CSIO
Chandigarh, India

Himanshu S. Moharana
Hi-Tech Institute of Technology
Khurda, Odisha, India

Qasim Murtaza
Delhi Technological University
New Delhi, India

G. L. Pahuja
National Institute of Technology
Kurukshetra, India

Tarun Panchal
CSIR-CSIO
Chandigarh, India

Inderpal Pasricha
Sri Guru Gobind Singh College
Chandigarh, India

Smruti Ranjan Pradhan
National Institute of Technical Teachers
    Training and Research
Chandigarh, India

Balwinder Raj
National Institute of Technical Teachers
    Training and Research
Chandigarh, India

P. Sudhakar Rao
National Institute of Technical Teachers
    Training and Research
Chandigarh, India

Mumtaz Rizwee
National Institute of Technical Teachers
    Training and Research
Chandigarh, India

Bharat Chandra Routara
School of Mechanical Engineering
KIIT (Deemed to be University)
Bhubaneswar, India

Susanta Kumar Sahoo
School of Mechanical Engineering
KIIT (Deemed to be University)
Bhubaneswar, India

Kanika Saini
National Institute of Technical Teachers
 Training and Research
Chandigarh, India

Stalin Kumar Samal
National Institute of Technical Teachers
 Training and Research
Chandigarh, India

Mantra Prasad Satpathy
School of Mechanical Engineering
KIIT (Deemed to be University)
Bhubaneswar, India

Mohd Bilal Naim Shaikh
Aligarh Muslim University
Aligarh, India

Anmoldeep Singh Sidhu
National Institute of Technical Teachers
 Training and Research
Chandigarh, India

Gurwinder Singh
National Institute of Technical Teachers
 Training and Research
Chandigarh, India

Manav Singh
JC Bose University of Science and
 Technology
YMCA
Faridabad, India

Rupinder Singh
National Institute of Technical Teachers
 Training and Research
Chandigarh, India

Surjeet Singh
U.I.E.T., Panjab University
Chandigarh, India

Yogesh Singh
National Institute of Technical Teachers
 Training and Research
Chandigarh, India

Ritula Thakur
National Institute of Technical Teachers
 Training and Research
Chandigarh, India

Vanraj
IAIS
Ambala Cantt., Ambala, India

Bodade Sandip Vasudeo
National Institute of Technical Teachers
 Training and Research
Chandigarh, India
and
Government Polytechnic
Jintur, Dist. Parbhani, India

Dinesh Chander Verma
Panipat Institute of Engineering and
 Technology
Panipat, India

Poonam Verma
Panipat Institute of Engineering and
 Technology
Panipat, India

Sanjeev Verma
Punjabi University
Patiala, Punjab

Section I

Computational Material Processing

1 On Dimensional Accuracy Modeling of Hot Chamber Die Casting

Rupinder Singh and Vinay Kumar

CONTENTS

1.1 INTRODUCTION

The hot chamber pressure die casting (HCPDC) machines are used primarily for low-melting-point alloys (zinc, copper, magnesium, lead, etc.) (Yoshihiko and Soichiro, 2009). It has been reported that in the HCPDC setup the furnace to melt the material is part of the die itself (Rosindale and Davey, 1999; Sabau and Dinwiddie, 2008; Okayasu et al., 2009). This process has been developed for achieving a high production rate and tight tolerance (Lee et al., 2006). The furnace used for melting metal/alloys is attached by a gooseneck. The process starts with rise of the injection cylinder plunger resulting in the opening of port in the cylinder. After this step, molten material is allowed to fill the cylinder. Further, with downward movement of plunger, molten matter is forced into the cavity. After solidification, the plunger is withdrawn, and the casting is ejected with the die opening (Cerri et al., 2008; Domkin et al., 2008; Kumar, 2010; Long et al., 2011). The reported literature outlined that some work has been reported on optimization of HCPDC like the effect of second phase pressure (Pr_2), metal pouring temperature (θ), and die opening time (T), etc. (Hallam and Griffiths, 2004; Dargusch et al., 2006; Singh and Kapoor, 2012; Singh and Singh, 2012). Figure 1.1 shows the web of processes that has been reported to study the casting of various industrial products. It can be ascertained from Figure 1.1 that a lot of advance manufacturing processes like high pressure casting, cold chamber die casting, etc., have been explored for the production of high-accuracy products. Figure 1.2 shows that little has been reported on

FIGURE 1.1 Web of terms investigated for study of advance manufacturing processes. (Based on VOS viewer software.)

the modelling of Δd in HCPDC. In the present case study, the spring adjuster (Figure 1.3) has been selected.

This macro model was developed for Δd in HCPDC. In the previous reported study, the effects of three process parameters (Pr_2, θ, and T) were revealed (Singh and Singh, 2012). This work is an extension of the previous study in which an empirical model based on the macro model has been prepared.

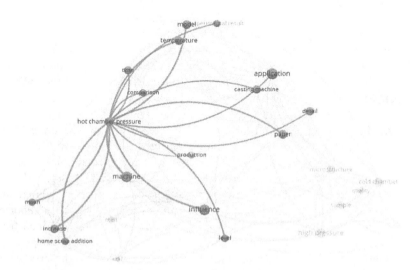

FIGURE 1.2 Areas least investigated for hot chamber pressure die casting. (Adapted from Figure 1.1.)

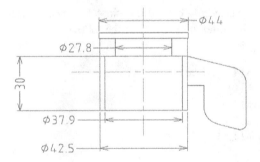

FIGURE 1.3 A two-dimensional design of component. (From Singh and Singh, 2012.)

1.2 MATHEMATICAL MODELING OF Δd IN HCPDC

As per previous reported studies, Δd in HCPDC is dependent on Pr_2, θ, and T. Tables 1.1 and 1.2 show the contribution of input parameters and a model for Δd (Singh and Singh, 2012).

TABLE 1.1
Percentage Contribution for Δd in HCPDC

	Sum Square	Percentage Contribution (% P)
Pr_2	1.1886747	1.5601844
θ	15.276193	20.050631
T	58.645386	76.974481
Error	1.0778351	1.4147029

TABLE 1.2
Geometric Model for Δd in HCPDC

	Optimized Conditions for Δd
Pr_2	19.61 N/mm²
Θ	420°C
T	5 s

Source: Singh and Singh (2012).

Buckingham's approach states that, in a physical problem having "a" quantities with "b" dimensions, the quantities may be arranged in to "a − b" dimensionless parameters (Singh, 2014; Singh and Singh, 2015). Based on the reported literature, Δd of HCPDC depends on Pr_2, θ, and T significantly and input parameters such as first phase pressure (Pr_1), limit switch position (LS), and clamping force (F) were not significant. Thus, by selecting basic dimensions M (mass), L (length), T (time), and θ, dimensions of other quantities would be the following:

Δd (mm)   L
Pr_1 (Nmm^{-2}) $M L^{-1} T^{-2}$
LS (cm) L^1
Pr_2 (Nmm^{-2}) $M L^{-1} T^{-2}$
θ (°C) θ
F (N) $M L^1 T^{-2}$
T (s)

Now, $\Delta d = f(Pr_1, Pr_2, LS, \theta, F, T)$

In this case a is 7 and b is 4. So (a − b = 3) and $¥_1$, $¥_2$, and $¥_3$ are three dimensionless groups. Taking Δd, Pr_2, and F as the quantities that directly go in $¥_1$, $¥_2$, and $¥_3$, respectively, it can be written in terms of Eqs. 1.1 to 1.3:

$$¥_1 = \Delta d \times LS^{\Psi 1} \times Pr_1{}^{\varphi 1} \times T^{\zeta 1} \times \theta^{\delta 1} \tag{1.1}$$

$$¥_2 = F \times LS^{\Psi 2} \times Pr_1{}^{\varphi 2} \times T^{\zeta 2} \times \theta^{\delta 2} \tag{1.2}$$

$$¥_3 = Pr_2 \times LS^{\Psi 3} \times Pr_1{}^{\varphi 3} \times T^{\zeta 3} \times \theta^{\delta 3} \tag{1.3}$$

Substituting the dimensions, the exponent of basic dimensions was achieved, since "$¥_{is}$" are dimensionless groups; thus, Ψi, φ, i, ζ_i, and δ_i, where i = 1, 2, 3 may be solved.
Solving for $¥_1$ (Eq. 1.1):

$$¥_1 = L \times L^{\Psi 1} \times (ML^{-1}T^{-2})^{\varphi 1} \times T^{\zeta 1} \times \theta^{\delta 1}$$

where
M: $\varphi_1 = 0$
L: $1 + \Psi_1 - \varphi_1 = 0$
T: $-2\varphi_1 + \zeta_1 = 0$
θ: $\delta_1 = 0$

Solving, we get:
$\Psi_1 = -1$, $\varphi_1 = 0$, $\zeta_1 = 0$, and $\delta_1 = 0$
Thus

$$¥_1 = \Delta d \times LS^{-1}$$

Similarly:

$$¥_2 = MLT^{-2} \times L^{\Psi 2} \times (ML^{-1}T^{-2})^{\varphi 2} \times T^{\zeta 2} \times \theta^{\delta 2}$$

where
M: $1 + \varphi_2 = 0$
L: $1 + \Psi_2 - \varphi_2 = 0$
T: $-2 - 2\varphi_2 + \zeta_2 = 0$
θ: $\delta_2 = 0$

Solving, we get:

$\Psi_2 = -2$, $\varphi_2 = -1$, $\zeta_2 = 0$, and $\delta_2 = 0$.

Thus, $\yen_2 = F. (LS)^{-2}. (Pr_1)^{-1}$

Also, $\yen_3 = M\,L^{-1}\,T^{-2} \times L^{\Psi_3} \times (M\,L^{-1}\,T^{-2})^{\varphi_3} \times T^{\zeta_3} \times \theta^{\delta_3}$

where

M: $1 + \varphi_3 = 0$

L: $-1 + \Psi_3 - \varphi_3 = 0$

T: is $-2 - 2\varphi_3 + \zeta_3 = 0$

θ: $\delta_3 = 0$

Solving, we get:

$\Psi_3 = 0$, $\varphi_3 = -1$, $\zeta_3 = 0$, and $\delta_3 = 0$

Thus

$$\yen_3 = Pr_2 (Pr_1)^{-1}$$

Since $\yen_i = f(\yen_j, \yen_k)$

$$\Delta d/(LS) = f\left(Pr_2 /Pr_1,\ F.\,(LS)^{-2}.\,(Pr_1)^{-1}\right)$$

$$\Delta d = f\left\{F.\,Pr_2\,/(LS).\,(Pr_1)^2\right\}$$

$$\Delta d = C\left\{F.\,Pr_2\,/(LS).\,(Pr_1)^2\right\}$$

Here C represents the HCPDC correction factor. By keeping F/LS $(Pr_1)^2$ as fixed, observations were made for different Pr_2 definitions to ascertain Δd and C. The experimental data for Δd with three different Pr_2 definitions have been plotted (Figure 1.4) for getting the best fitting curve. The equation for Δd may be written as $\Delta d = C \{F.\,Pr_2/(LS).\,(Pr_1)^2\}$:

- For Pr_2 14.71 N/mm^2 $\Delta d = (0.015\,Pr_2^2 - 0.125\,Pr_2 + 0.37)\,\{F/(LS).\,(Pr_1)^2\}$
- For Pr_2 19.61 N/mm^2 $\Delta d = (-0.01\,Pr_2^2 + 0.1\,Pr_2 - 0.17).\,\{F/(LS).\,(Pr_1)^2\}$
- For Pr_2 24.52 N/mm^2 $\Delta d = (0.04\,Pr_2^2 - 0.41\,Pr_2 + 1.14).\,\{F/(LS).\,(Pr_1)^2\}$

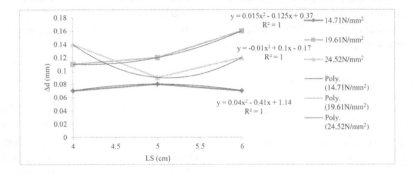

FIGURE 1.4   Graphical results for Δd vs LS.

By considering different parameters such as $Pr_2 = 14.71$ N/mm^2, F = 4 t, LS = 110 cm (1100 mm), and $Pr_1 = 90$ bar (9 N/mm^2):

$$\Delta d = \left[0.015\,(14.71)^2 - 0.125\,(14.71) + 0.37\right]\left\{4 \times 980.66\,/\,(1100).\,(9)^2\right\}$$

$$\Delta d = \left[3.24 - 1.83 + 0.37\right](0.044) = 0.07\,\text{mm}$$

The second-degree equation was observed after curve fitting with a coefficient of co-relation close to 1. These are in line with other investigators (Jufu et al., 2012; Lifang et al., 2012; Singh and Singh, 2015). For validation of this model, a comparison of Δd results was made for experimental and mathematical relations. The confirmatory experiment revealed that there is 37.7% improvement in Δd (Figure 1.5). For verification of model equations (as per Table 1.2) a historical data approach has been used. The empirical relation for Δd is given as $= -0.42 + 9.07 \times 10^{-4} \times Pr_2 + 10^{-3} \times \theta + 0.02 \times T$.

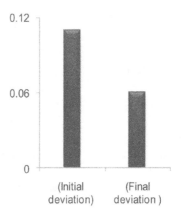

FIGURE 1.5   Improvement in Δd.

After using the data of optimized settings (Table 1.2), $\Delta d = 0.06$ mm; further similar results were observed under experimental investigations (Figure 1.6).

As observed from Figure 1.6, a normal distribution curve was followed for the selected data. The statistical analysis performed on the HCPDC component is also in line with these observations (Figure 1.7).

1.3   CONCLUSIONS

As per developed mathematical model after optimizing input parameters the Δd was observed as 0.06 mm. Further observations are in accordance with statistical analysis performed along radial and longitudinal dimensions. The developed polynomial equation expresses all significant input parameters as an extension of the macro model. It has been ascertained that the HCPDC process under model conditions will follow the normal distribution curve.

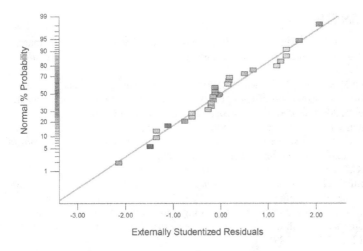

FIGURE 1.6 Experimental observations at optimized settings.

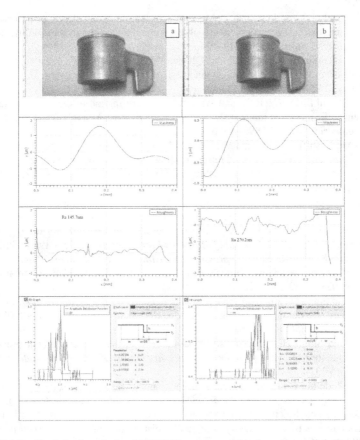

FIGURE 1.7 Statistical analysis of HCPDC component (a) along the radial dimension and (b) along the longitudinal direction.

REFERENCES

Cerri E., Leo P., and Marco P.P.D., Hot compression behavior of the AZ91 magnesium alloy produced by high pressure die casting, *Journal of Materials Processing Technology* 189(2008)97–106.

Dargusch M.S., Dour G., Schaeur N., Dinnis C.M., and Savaged G., The influence of pressure during solidification of high pressure die cast aluminium telecommunication components, *Journal of Materials Processing Technology* 180(2006)37–43.

Domkin K., Hattel J.H., and Thorborg J., Modeling of high temperature- and diffusion-controlled die soldering in aluminum high pressure die casting, *Journal of Material Processing Technology* 209(2008)4051–4061.

Hallam C.P., and Griffiths W.D., A model of interfacial heat transfer coefficient for the aluminium die casting process, *Metallurgical and Materials Transactions* 35B(2004)721–733.

Jufu J., Ying W., Yuanfa L., Jianjun Q., Weiwei S., and Shoujing L., A double control forming technology combining die casting and forging for the production of Mg alloy components with enhanced properties, *Journal of Materials Processing Technology* 212(2012)1191–1199.

Kumar L., Multi-response optimization of process parameters in cold chamber pressure die casting, Thesis, Department of Mechanical Engineering, Thapar University Patiala (2010).

Lee S.G., Patel G.R., and Gokhale A.M., Characterization of the effects of process parameters on macrosegregation in a high-pressure die-cast magnesium alloy, *Materials Characterization* 55(2006)219–224.

Lifang H., Shaoping C., Yang M., and Qingsen M., Die-casting effect on surface characteristics of thin-walled AZ91D magnesium components, *Applied Surface Science* 261(2012)851–856.

Long A., Thornhill D., Cecil C., and Watson D., Determination of the heat transfer coefficient at the metal die interface for high pressure die cast AlSi9Cu3Fe, *Applied Thermal Engineering* 31(2011)3996–4006.

Okayasu M., Yoshifuji S., Mizuno M., Hitomi M., and Yamazaki H., Comparison of mechanical properties of die cast aluminium alloys: cold vs. hot chamber die casting and high vs. low speed filling die casting, *International Journal of Cast Metals Research* 22(2009)374–381.

Rosindale I., and Davey K., Transient thermal model for the hot chamber injection system in the pressure die casting process, *Applied Mathematical Modeling* 23(1999)255–277.

Sabau A.S., and Dinwiddie R.B., Characterization of spray lubricants for the high pressure die casting processes, *Journal of Materials Processing Technology* 195(2008)267–274.

Singh R., and Kapoor R., *Effect of pressure on casting properties in cold chamber die casting*, Lambert Academic Publishing AG & Co. KG, Saarbrücken, Germany (2012), ISBN 978-3-659-12721-2.

Singh R., and Singh H., *Effect of some parameters on properties of hot chamber die casting*, Lambert Academic Publishing AG & Co. KG, Saarbrücken, Germany (2012), ISBN 978-3-659-13087-8.

Singh R., and Singh H., Effect of some parameters on the cast component properties in hot chamber die casting, *Journal of the Institution of Engineers (India): Series C* 97(2015)131–139.

Singh R., Modeling of surface hardness in hot chamber die casting using Buckingham's pi approach, *Journal of Mechanical Science and Technology* 28(2014)699–704.

Yoshihiko H., and Soichiro K., Quantitative evaluation of porosity in aluminium alloy dies castings by fractal analysis of spatial distribution of area, *Material and Design* 30(2009)1169–1173.

2 Thermomechanical Properties and Surface Morphology of MMT-Neoprene Nanocomposites

Mohd. Gulam Waris Khan and Abdul Faheem

CONTENTS

2.1 INTRODUCTION: RESEARCH BACKGROUND

Nanocomposite technology has attracted a great deal of attention in materials science. Remarkable importance has been bestowed upon polymer nanocomposites over the last few decades. Recently high-performance nanocomposites based on organic polymer/inorganic fillers have provided a new platform to the investigators in materials science because of their unique and excellent properties. The main challenge lies in the efficient dispersion and alignment of nanofiller into the polymeric matrix. In the early days, rubber materials stayed strengthened with nanofillers to enhance their stability and durability by integrating normal fillers. Polychloroprene (neoprene) rubber (CR) comes under the label of special-purpose rubbers broadly applicable in the field of electromechanical sectors. This rubber is used in specific applications that require excellent physical toughness, solvent resistance, thermal resistance, resistance to inflammation in mineral, animal, and plant oils, fats, etc. Neoprene rubber goods include seals, hoses, profiles, rolls, belts, V-belts, bearings, linings, rubberized fabrics, shoe soling, and many applications in the construction industry. Nanocomposite technology has attracted a great deal of attention in materials science. Now polymer nanoclay composites are more interesting to researchers and scientists because they exhibit numerous industrial uses. Materials through a combination of nanoparticle-sized inorganic and organic materials with polymeric

DOI: 10.1201/9781003203681-3

11

materials that provide characteristics and properties of the single constituents by reinforcing constituents are nanoclay, nano silica, nano graphite, and single-wall and multiwall carbon nanotubes (MWCNTs). Polymer nanocomposites are those materials with reinforcements that are homogeneously dispersed in nanometer measures inside the polymeric matrix (Morrill, 1987; Mark, 1996; Paul et al., 2003; Nguyen and Baird, 2006; Sisanth et al., 2017; Sivaraman et al., 2017).

Yeha et al. (2007) investigated the morphological structure and mechanical strength of neoprene rubber and montmorillonite (MMT) nanoclay-based nanocomposites and modified MMT nanoclay was used. The X-ray diffractometer was used to determine the corresponding modification in the D spacing. Scanning electron microscopy (SEM) was used to examine the morphological behavior of several reinforced composites. Transmission electron microscopy (TEM) was used to examine the reinforcement of MMT nanoclay and neoprene rubber. The consequences show that the incorporation of MMT nanoclay improves the mechanical stability and strength of neoprene rubber. Rajkumar et al. (2013) and Rajkumar (2014) stated the influence of dioctyl phthalate as the medium of dispersion on the MMT nanoclay and nitrile rubber (NBR) matrix and dispersal of nano silica in the CR matrix. Nanoclay in CR enhanced mechanical strength and good dispersal by enhancing nanoclay filler. The resulting polymer comprises nanoclay fillers entrenched in the medium polymerization that can be cross-linked to achieve the vulcanized elastomer nanocomposite. Nanocomposites prepared with nanoclay fillers were revealed to provide outstanding property improvements equating to normal microcomposites and on normal fillers. Singh et al. (2012) have studied the development of CR composites reinforced with MWCNTs. It has been observed that the incorporation of 0.3 wt.% of filler in the elastomer matrix indicates an important enhancement in the physicomechanical strength of the nanocomposites.

Today's MMT nanoclay is the best nanofiller that is applicable for the production of nanocomposite materials because it has decent thermomechanical properties. The required nanoclay filler base material is an MMT-layered smectite clay mineral that has a platypoecilus structure. The depth of a single platelet is 1 nm and surface dimensions are normally 300–600 nm consequential to a remarkably high strength-weight ratio. Subsequently, polymeric materials are commonly organophilic, and unchanged nanoclay diffuses in the polymeric matrix with trouble. Finished nanoclay surface reforms, and MMT can be made organophilic and, consequently, compatible with normal organic polymeric materials. Surface compatibility is well known as "intercalation." The compatibilized nanoclay diffuses readily in the polymeric structure (Zhang et al., 2003; Liang et al., 2004; Tambe et al., 2009; Bee et al., 2018).

CR is a rubber commonly used in electrical and mechanical industries. It has outstanding toughness; a higher working range of thermal comparatively hydrocarbon-based rubber; and excellent resistance to sunlight, ozone (O_3), and normal climate situations. The stability of these constituents is essentially dependent on the tensile properties of CR rubber.

In the present study, an attempt has been made through MMT-based neoprene nanocomposite and to study their thermomechanical properties with surface morphology.

2.2 EXPERIMENTAL APPROACH

In this chapter, the following ingredients were used: CR obtained from DuPont USA grade W-M1; zinc oxide; stearic acid; 2,6-ditert-butyl-4-methylphenol; 2-mercapto-benzothiazole (MBT); dicumyl peroxide; thiourea, which was procured from E. Merck Germany; di-octyl-phthalate obtained from S. R. Chemical Kanpur; MMTs called organo-modified montmorillonite (OMMT) clay with nanomer 1.30 PS; and MMT clay surface modified with 20–340 wt.% of octadecylamine. In addition, 0.5–5 wt.% aminopropyltriethoxysilane was used as reinforcement for the preparation of nano-composites supplied from Sigma Aldrich (USA).

In a traditional laboratory open to two-roll mills (150 × 330 mm), compounding was done at 140°C–150°C, and neoprene elastomer was blended with other ingredi-ents as shown in Table 2.1. The compounded elastomer has been molded to obtain 3.5-mm sheet thickness using an electrically heated hydraulic compression molding press at 150°C for 20 minute at a pressure of 2400 psi. These cured sheets have to be conditioned before testing (24-hour maturation at 25°C). The samples were prepared through sheets of neoprene nanocomposites with compositions as per ASTM-D standards. These samples were utilized for evaluating the thermal and mechanical performance of the nanocomposite. Before testing, the working samples were condi-tioned at 23 ± 2°C with 50 ± 5% relative humidity for 24 hours.

Different properties, i.e., tensile modulus and elongation, are evaluated under ASTM D638 (dumbbell) shaped samples. The test was performed by INSTRON UTM accompanied by model 3382 at room temperature with a gauge length and crosshead speed of 35 mm and 5 mm/min, respectively. Tensile values reported here are an average of the result for a test run on at least four specimens. The hardness of the nanocomposites was measured under ASTM D2240 by Shore A durometer. Pyris TGA-I (PerkinElmer, USA) thermal analyzer was used for the thermogravimetric analysis (TGA) that gave the degradation, pattern, and thermal stability of the nano-composites. Under an inert atmosphere, the sample weighed about 4 mg and heated at the range of 50°C–650°C at the controlled rate of heating at 10°C/min. SEM was also investigated with Zeiss EVO-50 VP for the characterization of mechanical and surface morphology.

TABLE 2.1
Formulation of the Nanocomposite Based on CR Rubber

S. code	Neoprene Rubber (wt.%)	Formulation (phr)							
		ZnO	Stearic Acid	Antioxidant	Thiourea	MBT	DCP	DOP	MMT
A	100	5	2	1.75	2	1.5	2	10	0
B	100	5	2	1.75	2	1.5	2	10	1
C	100	5	2	1.75	2	1.5	2	10	3
D	100	5	2	1.75	2	1.5	2	10	5

Abbreviations: DCP, dicumyl peroxide; DOP, dioctyl Phthalate; MMT, montmorillonite.

2.3   RESULTS AND DISCUSSION

The results in Table 2.2 depict a significant enhancement in the tensile strength of the neoprene/MMT nanoclay nanocomposite. The use of MMTs in the neoprene matrix contributes to a larger increase in tensile strength. The higher weight to strength ratio of the MMT nanocomposite may be due to its reinforcement. Tensile modulus and elongation at rupture point increases have also been observed. This may be because of the direct bonding of MMTs and polymer chains, thus, providing better wetting or adhesion at the interface of two phases. The hardness of neoprene/MMT has been increased from 1 to 5 parts per hundred (phr) loading of MMT nanoclay.

The results in Table 2.3 TGA studies have been conducted for various loadings of coated MMT in CR matrix. TGA thermograms (Figure 2.1) depict that for all the nanocomposites there is a huge enhancement in the thermal characteristics as contrasted with the unfilled CR matrix. This might be due to the reduction of chain mobility of the CR matrix by providing a large number of restriction sites, resulting in reducing the thermal vibration of active bonds. Thus, we can say that developed nanocomposites require much more thermal energy for the decomposition of CR matrices, which enhances the thermal stability. It is also important to note that the degradation temperature of CR rubber continually increases with rising MMT content and reaches the maximum value of 5 wt.% of the MMT nanoclay in the neoprene elastomer matrix.

TABLE 2.2
Mechanical Properties of the CR Elastomer Reinforced with MMT

Sample	Tensile Strength (MPa)	Tensile Modulus (MPa)	Elongation at Rupture	Hardness
A	0.98	1.1	454	60
B	1.29	1.2	493	64
C	1.53	1.5	565	67
D	1.81	1.8	678	67

TABLE 2.3
TGA for Different Samples

Description	Loading MMT (%)	Weight Loss (%)	Residue After 810°C
CR		66	33.476
CR/MMT	1	67	32.103
CR/MMT	3	69	30.990
CR/MMT	5	68	31.068

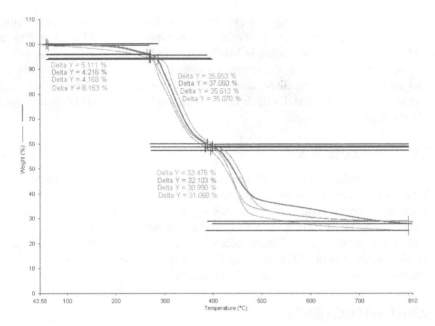

FIGURE 2.1   TGA analysis of MMT nanoclay.

The characterization approach of SEM depicts that pure neoprene vulcanized has a very smooth surface with some small embedded particles as in Figure 2.2(a). MMT-filled vulcanizates exhibit a rougher fractured surface with the unusual ridge-lines, as can be seen in Figure 2.2(b). The particles visualized in the virgin CR matrices are probably the curvatures used in vulcanizates. The SEM micrographs as

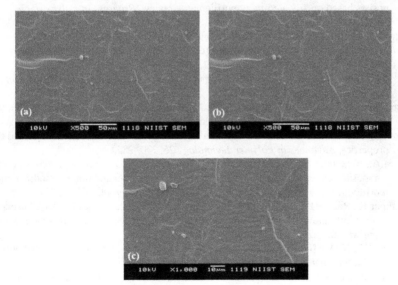

FIGURE 2.2   Scanned electron microscopy.

depicted in Figure 2.2(c) exhibit an excellent dispersion of the MMT, which results in a more homogeneous network. From the observations of SEM micrographs, it can be concluded that the loading of MMT results in better dispersion of the MMT. Significant enhancement in the tensile properties of the nanocomposites may be attributed to the excellent dispersion of the MMT in the CR matrix. It has been observed that nanofiller and a thick layer of CR appear in the MMT, indicating a salutary degree of adhesion.

2.4   CONCLUSIONS

In this study, it has been demonstrated that there is a drastic enhancement in the thermomechanical properties and surface morphology. Mechanical properties, like tensile properties, such as strength, modulus, and elongation, enhance as the loading of MMT nanoclay is increased. (at 5 phr of MMT). Higher loading of coated MMT in the CR matrix depicts a higher value of tensile strength, thermal stability, and excellent dispersion of MMT nanoclay in the rubber matrix. The developed nanocomposites may find applications in the defense and civil sectors.

ACKNOWLEDGMENTS

The authors are thankful to CIPET, Lucknow, for experimental facilities, and the Mechanical Engineering Section, University Polytechnic, AMU, Aligarh, India, for their kind support and guidance.

REFERENCES

Bee S.L., Abdullah M.A.A., Bee S.T., Sinc L.T., and Rahmat A.R., 2018. Polymer nanocomposites based on silylated-montmorillonite: a review, *Progress in Polymer Science* 85, 57–82.

Liang G., Xu J., Bao S., and Xu W., 2004. Polyethylene/maleic anhydride grafted polyethylene/organic-montmorillonite nanocomposites. Preparation, microstructure, and mechanical properties, *Applied Polymer Science* 91, 3974–3980.

Mark J.E., 1996. Ceramic-reinforced polymers and polymer-modified ceramics, *Polymer Engineering Science* 36, 2905–2920.

Morrill J.P., 1987. Nitrile and polyacryalate rubbers. In: Morton, M. ed., *Rubber Technology*, 3rd ed. Maurice Van Nostrand Reinhold, New York, p. 638.

Nguyen Q.T., and Baird D.G., 2006. Preparation of polymer–clay nanocomposites and their properties, *Advances in Polymer Technology* 25, 270–285.

Paul M.A., Alexander M., Degee P., Henrist C., Rulmont A., and Dubois P., 2003. New nanocomposites materials based on plasticized poly (L-lactide) and organo-modified montmorillonites, *Thermal and Morphological Study, Polymer* 44, 443–450.

Rajkumar K., Ranjan P., Thavamani P., Ponnusamy J., and Pazhanisamy P., 2013. Dispersion studies of nanosilica in NBR based polymer nanocomposite, *Rasayan Journal Chemistry* 6, 122–133.

Rajkumar K., 2014. Effect of DOP as dispersion medium on MMT-nano clay in NBR polymer matrix, *Indian Journal of Applied Research* 4, 29–32.

Singh Y.P., Pandey K.N., Verma V., Singh P., and Mishra R.M., 2012. *Thermal, mechanical and microstructure properties of neoprene-MWCNT nanocomposites*, Thesis, Department of Engineering, CIPET Lucknow, India.

Sisanth K.S., Thomas M.G., Abraham J.S., Pal K., Panwar V., and Bahadur J., 2017. *General introduction to rubber compounding and rubber blend nanocomposites.* Woodhead Publishing Series in Composites Science and Engineering. Woodhead Publishing, Cambridge, UK.

Sivaraman O., Ghosh N., Gayathri S., Sudhakara P., Misra S.K., and Jayaramudu J., 2017. *Natural rubber nanoblends: preparation, characterization and applications.* Springer Series on Polymer and Composite Materials. Springer Nature, Switzerland.

Tambe S.P., Naik R.S., Singh S.K., Patri M., and Kumar D., 2009. Studies on effect of nano-clay on the properties of thermally sprayable EVA and EVAI coatings, *Progress in Organic Coatings* 65, 484–489.

Yeha M.H., Hwanga W.S., and Chengb L.R., 2007. Microstructure and mechanical properties of neoprene–montmorillonite nanocomposites, *Applied Surface Science* 253, 4777–4781.

Zhang Q.X., Yu Z.Z., Yang M., Ma J., and Mai Y. W., 2003. Multiple melting and crystallization of nylon-66/montmorillonite nanocomposites, *Journal of Polymer Science Polymer Physics* 41, 2861–2870.

3 An Overview of Friction Stir Welding Tools

Sanjeev Verma and Vinod Kumar

CONTENTS

3.1 INTRODUCTION

Friction stir welding (FSW) was invented at The Welding Institute (TWI) in the United Kingdom in 1991. FSW is an appropriate welding process used for the joining of different combinations of dissimilar aluminum alloys (Murr, 2010). In the FSW process, a rotating tool of nonconsumable material harder than the base material is used to join the workpiece. Tool rotation with its movement to the joint line of the workpiece plastically deforms the surrounding material, which generates a great deal of heat between the base material and tool shoulder resulting in the base materials being welded during the process (Mishra and Ma, 2005; Elangovan and Balasubramanian, 2008). The basic diagram for the FSW process is illustrated in Figure 3.1 in which the tool rotation and welding direction are shown. The tool's downward force acts during the process and the tool pin is inserted into the workpiece, which generates the heat that leads to displacement of material and stirring the materials between each other. Due to the stirring action, the materials are joined in their solid state (Ghaffarpour et al., 2012). The conventional fusion welding technique used for different materials and for aluminum alloys results in various welding defects like hot cracking, voids, lack of penetration, etc., in welded joints (Palanivel et al., 2012; Verma and Pandey, 2012). Thus, a new solid state welding technique (i.e., FSW) is widely employed to solve these types of defect problems, which result in fewer defects (Huskins et al., 2010). The high strength, absence of melt-related defects, and low distortion makes FSW better than conventional fusion welding techniques.

DOI: 10.1201/9781003203681-4

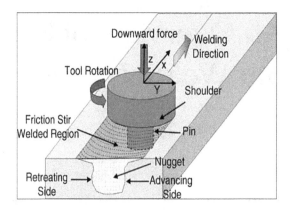

FIGURE 3.1   FSW process. (From Palanivel et al., 2012 and Bagheri et al., 2013.)

3.2 FUNCTION OF FSW TOOL

Localized heating and material flow are the two primary functions of the FSW tool. The frictional heat is generated due to the contact between the tool pin and workpiece, resulting in deformation and softening of material, which creates a joint in workpiece. In the initial functional stage, the FSW tool pin is plunged until the tool shoulder touches the workpiece face, which results in heat formation. Another function of the tool is to create a stirring action and move the stirred material during the process, which depends on the design of tool (Mishra and Ma, 2005).

3.3 TOOL MATERIAL

The properties of the tool material affect the joint quality by influencing heat dissipation and heat generation during the welding process. Tool material should be of harder material compared with the base material of the workpiece. Weld quality, high operating temperature range, and tool wear resistance are important factors for selection of tool material. It is always undesirable if a tool loses dimension stability, causes fractures, or worse (Mishra and Mahoney, 2007). The other factors that influence the tool material are ductility, strength, hardness, and reactivity with the workpiece. Tool material should have the following characteristics:

- Ambient and elevated temperature strength with stability
- Good wear resistance and high fracture toughness
- Minimum or no reactivity
- Good machinability property

The most commonly used tool materials with their properties are mentioned in the following:

- *Hot-worked tool steel materials:* Good machinability, wear resistant, and good thermal fatigue resistance; suitable for welding of aluminum alloys and copper materials

- *Nickel and cobalt base alloys:* Good creep resistance, excellent ductility, have high strength
- *Refractory materials (W, Mo):* Suitable for strong alloys with a temperature range of 1000–1500°C, brittle in nature, high temperature stability
- *Tungsten-based alloys:* Have high cost compared with other tool materials, good strength, high temperature stability
- *Metal matrix composites of carbide material (WC, WC-Co, TiC):* Good stability at high temperature operation, excellent wear resistance, very expensive tool material, good fracture toughness (Meilinger and Torok, 2013)

3.4 TOOL GEOMETRY

Tool geometry is the most critical factor for the fabrication of sound welds. Tool geometry influences the heat generation during the welding process, the transverse force, and the thermomechanically environment experienced by the tool. The shoulder-to-pin diameter plays an important role in the stir zone and should be considered very accurately so that sound welds can be achieved (Mishra and Ma, 2005). Tool geometry depends on the shoulder diameter and the tool pin profile and its shape (Colegrove and Shercliff, 2004; Mishra and Ma, 2005; Fujii et al., 2006; Hattingh et al., 2008; Badarinarayan et al., 2009; Sorensen and Nielsen, 2009; Sun et al., 2009; Hirasawa et al., 2010; Lorrain et al., 2010; Vijay and Murugan, 2010).

3.4.1 DESIGN OF TOOL SHOULDER DIAMETER

The tool shoulder design includes its diameter, and length is an essential factor for generation of heat on the surface and subsurface region of the workpiece. The sticking and sliding factors are responsible for heat generation, whereas sticking is responsible only for material flow during operation. The optimal value of the shoulder diameter results in the highest strength of welded joints. Figure 3.2 shows a relationship diagram between tool shoulder and peak temperature for rotational speeds of 344 rpm and 560 rpm, and illustrates a comparison between computed and measured values (Mehta et al., 2011).

From this comparison diagram, it was concluded that with an increase in tool (shoulder) diameter and in rotational speed, peak temperature increases. Larger shoulder diameters exhibit larger contact area, which results in higher frictional heat that leads to higher temperature. Tool shoulder diameter shape is also an important consideration. Basically three shoulder diameter shape designs are used, which influence the weld quality of the joint. The widely used shoulder profiles are flat surface (Zhang et al., 2011; Sabari et al., 2016), concave (Wang et al., 2015), and convex shape designs for FSW. All types of shoulder designs are illustrated in Figure 3.3(a). The simplest design of tool shoulder is flat and cannot control the flowing material during the process. The most popular concave design shoulder produces quality welds compared with other shoulder shapes. The concave design helps the material to flow to the center due to its shape and produce sound welds. The convex shape tool shoulder displaces the materials away from the probe so that proper mixing cannot be done. The bottom surface of the shoulder is also important as it increases

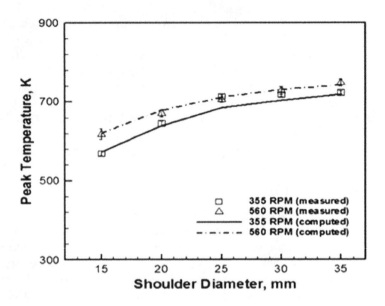

FIGURE 3.2 Effect of shoulder diameter and rotational speed on peak temperature. (From Mehta et al., 2011.)

frictional and material deformation that enhances the mixing of material (Wang et al., 2015). Figure 3.3(b) shows some bottom surfaces of the shoulder. The proper mixing of material depends on the shape of the bottom surface of the shoulder.

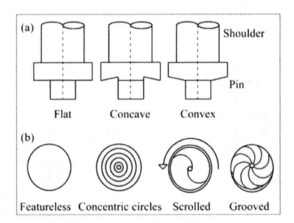

FIGURE 3.3 (a) Tool shoulder profile and (b) Bottom surface of shoulder. (From Wahid et al., 2018.)

3.4.2 TOOL PIN (PROBE) GEOMETRY

The researchers' theories show that tool geometry affects the weld properties and material flow during the welding process (Colegrove and Shercliff, 2004; Mishra and

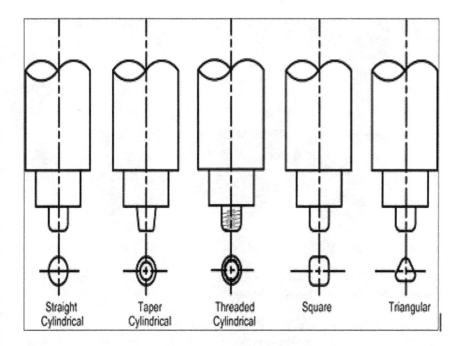

FIGURE 3.4   Commonly used pin (probe) profiles for FSW.

Ma, 2005; Nandan et al., 2008; Choi et al., 2010; Lorrain et al., 2010; Yadava et al., 2010). There are various types of pin profiles used for the FSW process. The tool pin profile depends on the application for which it is used; each pin profile has its own characteristics to produce sufficient heat as well as material flow control during the process. The most commonly used pins (probes) are shown in Figure 3.4.

Other important tool pin geometries used for FSW of dissimilar alloys are illustrated in Figure 3.5. These include the following:

- *Round-bottom cylindrical pin (Figure 3.5(a)):* The advantage of the round-bottom cylindrical pin is that its pin reduces the wear rate of the tool during plunging so that the weld quality is improved.
- *Flat-bottom cylindrical pin (Figure 3.5(b)):* The frictional velocity of this type of tool pin is zero in the center and maximum at the cylinder's edge.
- *Cylindrical pin shapes(Figure 3.4):* These are used for the aluminum alloys up to a 12-mm thickness. Other modified shapes are square, triangular, taper cylindrical, hexagonal, and threaded pin shape.
- *MX triflute pin profile (Figure 3.5(c)):* The thick sections of aluminum alloys can be welded by using the MX triflute pin shape tool as it increases the deformed material to the weld line as well as increases the tool travel speed due to the three flutes on its phase.

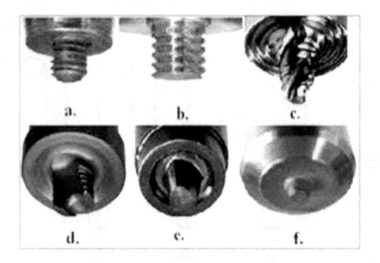

FIGURE 3.5 Different tool pin geometries: (a) round-bottom, (b) flat-bottom, (c) MX tri-flute, (d) A-skew™, (e) Trivex, and (f) threadless.

- *A-skew™ (Figure 3.5(d)):* This type of pin profile improves tensile properties of welded specimens, increases travel speed, and reduces the weld asymmetry.
- *Trivex pin (Figure 3.5(e)):* This pin profile shape reduces the transvering forces.
- *Threadless pin profile (Figure 3.5(f)):* It is used in special applications of FSW; it consists of a simple design with robust characteristics.

3.5 TOOL WEAR AND FAILURE

The wear properties of a tool are important for tool design, and the wear that occurs in the tool is due to tool translation as well as tool rotation movement during the process. In a high-load environment at elevated temperature, the welding tool deforms due to yield strength, so in some cases for high-strength materials the welding tool may often be liquid cooled during the process (Cam, 2011). The reaction of the tool material with the workpiece as well as with its environment may also be the reason for tool wear. Tool failure may also be due to the stresses that occur during the process or if they are higher than its load bearing capacity. The tool can wear out after making a number of joints. In case of aluminum or other light materials such as Mg alloys, tools made of steel are used that show very little wear. There is no significant wear reported on steel tools from Al alloy welding. High-strength materials such as steel, Ti, Ni, etc., are suitably welded by tools made of hard metals like carbides or MMC, which have high thermal and wear resistance. Figure 3.6 shows the tool wear of tool steel used for FSW of Al 6061 + 20 vol% Al_2O_3-MMC.

It has been reported that with increasing of weld speed, the tool wear considerably decreased. However, it was concluded that a decline was observed in tool wear after initial wear and then it smoothed (or self-optimized) the tools (shown in Figure 3.6)

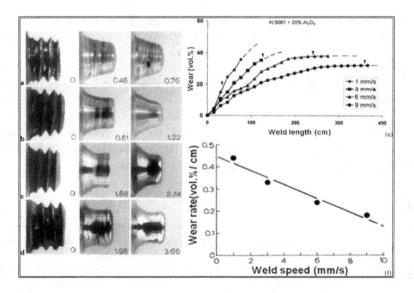

FIGURE 3.6 Tool wear features for Al 6061 + 20 vol% Al$_2$O$_3$-MMC at 1000 rpm: rotational speeds of (a) 1, (b) 3, (c) 6, and (d) 9 mm s^{-1}; (e) pin tool wear (vol%) versus weld length (cm); and (f) pin tool wear rate (vol%) versus weld speed (mm/s). (From Prado et al., 2003.)

that fabricate the sound joints (Shindo et al., 2002; Prado et al., 2003; Fernandez and Murr, 2004).

3.6 CONCLUSIONS

Many designing factors are involved in the fabrication of FSW tools. The tool designer must pay attention to the design of the FSW tool by considering the tool geometry, material, quality, and cost of each tool. The main conclusions are summarized in the following:

1. Various mechanical properties of tools like strength, thermal conductivity, thermal expansion, fracture toughness, and hardness highly affect the tool wear and weld quality of joints. The reactivity with oxygen and with the workpiece material are important considerations for designing the tool.
2. Sticking and sliding are responsible for heat generation, whereas sticking is only responsible for material flow during operation.
3. By increasing values in shoulder diameter and rotational speed, peak temperature increases. Larger shoulder diameters exhibit larger contact areas, which results in a higher frictional heat that leads to higher temperature.
4. The convex shape tool shoulder displaces the materials away from the probe, so that proper mixing cannot be done. The bottom surface of the shoulder is also important as it increases frictional and material deformation, which enhance the mixing of the material. The tool pin profile and tool shoulder diameter are responsible for heat generation and plastic flow of material.

5. The unique properties of hardness, high temperature stability, and high strength of pcBN tools make them better than other tools for FSW for high-strength harder materials.
6. Tool wear is an important consideration for designing tools. The tool wear is also affected by the weld speed; as the weld speed increases, the tool wear decreases.

In the field of the FSW process, there is a need to develop reliable and cost-effective tools for hard materials and there is a future demand for high-strength, hard, and high-melting-temperature tools for commercial applications.

REFERENCES

Badarinarayan H., Yang Q., and Zhua S., 2009. Effect of tool geometry on static strength of friction stir spot-welded aluminum alloy, *International Journal of Machine Tools and Manufacture* 49(2), 142–148.

Bagheri A., Azdast T., and Doniavi A., 2013. An experimental study on mechanical properties of friction stir welded ABS sheets, *Materials & Design* 43, 402–409.

Cam G., 2011. Friction stir welded structural materials: beyond Al- alloys, *International Materials Reviews* 56(1), 1–48.

Choi D.H., Ahn B.W., Lee C.Y., Yeon Y.M., Song K., and Jung S.B., 2010. Effect of pin shapes on joint characteristics of friction stir spot welded AA5J32 sheet, *Materials Transactions* 51(5), 1028–1032.

Colegrove P.A., and Shercliff H.R., 2004. Two-dimensional CFD modelling of flow round profiled FSW tooling, *Science and Technology of Welding and Joining* 9(6), 483–492.

Elangovan K., and Balasubramanian V., 2008. Influences of tool pin profile and tool shoulder diameter on the formation of friction stir processing zone in AA6061 aluminium alloy, *Materials & Design* 29, 362–373.

Fernandez G.J., and Murr L.E., 2004. Characterization of tool wear and weld optimization in the friction-stir welding of cast aluminum 359z20% SiC metal-matrix composite, *Materials Characterization* 52(1), 65–75.

Fujii H., Cui L., Maeda M., and Nogi K., 2006. Effect of tool shape on mechanical properties and microstructure of friction stir welded aluminum alloys, *Materials Science and Engineering A* A419(1–2), 25–31.

Ghaffarpour M., Dariani B.M., Kokabi A.H., and Razani N.A., 2012. Friction stir welding parameters optimization of heterogeneous tailored welded blank sheets of aluminium alloys 6061 and 5083 using response surface methodology, *Proceedings of the Institution of Mechanical Engineers, Part B: Journal of Engineering Manufacture* 226, 2013–2022.

Hattingh D.G., Blignault C., Niekerk T.I., and James M.N., 2008. Characterization of the influences of FSW tool geometry on welding forces and weld tensile strength using an instrumented tool, *Journal of Materials Processing Technology* 203(1–3), 46–57.

Hirasawa S., Badarinarayan H., Okamoto K., Tomimura T., and Kawanami T., 2010. Analysis of effect of tool geometry on plastic flow during friction stir spot welding using particle method, *Journal of Materials Processing Technology* 210(11), 1455–1463.

Huskins E.L., Cao B., and Ramesh K.T., 2010. Strengthening mechanisms in an Al–Mg alloy, *Materials Science and Engineering A* 527, 1292–1298.

Lorrain O., Favier V., Zahrouni H., and Lawrjaniec D., 2010. Understanding the material flow path of friction stir welding process using unthreaded tools, *Journal of Materials Processing Technology* 210(4), 603–609.

Mehta M., Arora A., De A., and Debroy T., 2011. Tool geometry for friction stir welding - optimum shoulder diameter, *Metallurgical and Materials Transactions A* 42A, 2716.

Meilinger A., and Torok I., 2013. The importance of friction stir welding tool, *Production Processes and Systems* 6, 25–34.

Mishra R.S., and Ma Z.Y., 2005. Friction stir welding and processing. *Materials Science and Engineering R: Reports*, 50, 1–78.

Mishra R.S., and Mahoney M.W., 2007, *Friction Stir Welding and Processing*. ASM International, Materials Park, OH, p. 6–19.

Murr L.E., 2010. A review of FSW research on dissimilar metal and alloy systems, *Journal of Materials Engineering and Performance* 19, 1071–89.

Nandan R., DebRoy T., and Bhadeshia H., 2008. Recent advances in friction-stir welding – process, weldment structure and properties, *Progress in Materials Science* 53(6), 980–1023.

Palanivel R., Mathews P.K., Murugan N., and Dinaharan I., 2012. Effect of tool rotational speed and pin profile on microstructure and tensile strength of dissimilar friction stir welded AA5083-H111 and AA6351-T6 aluminum alloys, *Materials &Design* 40, 7–16.

Prado R.A., Murr L.E., Soto K.F., and McClure J.C., 2003. Self-optimization in tool wear for friction-stir welding of Al 6061z20% Al2O3 MMC, *Materials Science and Engineering: A* A349(1–2), 156–165.

Sabari S.S., Malarvizhi S., Balasubramanian V., and Reddy G., 2016. The effect of pin profiles on the microstructure and mechanical properties of underwater friction stir welded AA2519-T87 aluminium alloy, *International Journal of Mechanical and Materials Engineering* 11(5), 1–14.

Shindo D.J., Rivera A.R., and Murr L.E., 2002. Shape optimization for tool wear in the friction-stir welding of cast Al359-20% SiC MMC, *Journal of Materials Science* 37(23), 4999–5005.

Sorensen C., and Nielsen B., 2009. Exploring geometry effects for convex scrolled shoulder, step spiral probe FSW tools, Proceedings of the TMS 2009 Annual Meeting, San Francisco, CA, p. 85–92.

Sun N., Yin Y.H., Gerlich A.P., and North T.H., 2009. Tool design and stir zone grain size in AZ31 friction stir spot welds, *Science and Technology of Welding and Joining* 14(8), 747–752.

Verma R.P., and Pandey K.N., 2012, Investigation of fatigue life of 6061-T6 and 5083-O aluminium alloys welded by two welding processes – manual metal arc welding and metal inert gas welding, International Conference on Mechanical and Industrial Engineering (ICMIE), p. 46–50.

Vijay S.J., and Murugan N., 2010. Influence of tool pin profile on the metallurgical and mechanical properties of friction stir welded Al– 10 wt.% TiB2 metal matrix composite, *Materials & Design* 31(7), 3585–3589.

Wahid M.A., Khan Z.A., and Siddiquee A.N., 2018. Review on underwater friction stir welding: A variant of friction stir welding with great potential of improving joint properties, *Transactions of Nonferrous Metals Society of China* 28, 193–219.

Wang Q., Zhao Z., Zhao Y., Yan K., and Zhang H., 2015. The adjustment strategy of welding parameters for spray formed 7055 aluminum alloy underwater friction stir welding joint, *Materials & Design*, 88, 1366–1376

Yadava M.K., Mishra R.S., Chen Y.L., Carlson B., and Grant G.J., 2010. Study of friction stir joining of thin aluminium sheets in lap joint configuration, *Science and Technology of Welding and Joining* 15(1), 70–75.

Zhang H.J., Liu H.J., and Yu, L., 2011. Microstructure and mechanical properties as a function of rotation speed in underwater friction stir welded aluminum alloy joints, *Materials & Design* 32, 4402–4407.

4 Ultrafast Laser for Processing of Materials

An Industry Perspective

Himanshu S. Moharana, Smruti Ranjan Pradhan, and Abhijeet Gupta

CONTENTS

4.1 INTRODUCTION: BACKGROUND AND DRIVING FORCES

Over the last decade, the processing of materials by ultrashort laser pulses has steadily grown and is beginning to render its scientific, technological, and industrial fields. In a broader variety of fields related to academic and research in engineering, the possibility of three-dimensional (3D) writing in glass (Davis et al., 1996) and polymers (Maruo et al., 1997) using closely focused femtosecond (fs) pulses of laser (discovered around 20 years ago) has gained high praise. The development of these objects with a size comparable to a living cell and even finer (Kawata et al., 2001) suggests the use of remotely controllable 3D microbots for healing missions and optical data processors incorporated into structures of robust optical memory that are not erasable. Thus, the race toward these and many other appealing goals began, and most of these goals have yet to be accomplished. Progress made in certain areas has been notable. Data density on optical memories with greater than 1 TBIT cm^3 (Glezer et al., 1996; Watanabe et al., 1998), optical information processing structures based on waveguides, optical quantum computing systems components (Marshall et al., 2009; Malinauskas et al., 2015), 3D photonic crystals (PhCs) (Mikutis et al., 2013), and micromechanical/biological systems (Malinauskas et al., 2014) can currently be obtained using ultrashort laser pulses. A modest discussion in this work focuses on accomplishments and discusses current developments in growth of laser processing, which is bound to make ultrashort laser production useable for nanotechnology applications. Different parameters affecting micromachining of metals

DOI: 10.1201/9781003203681-5

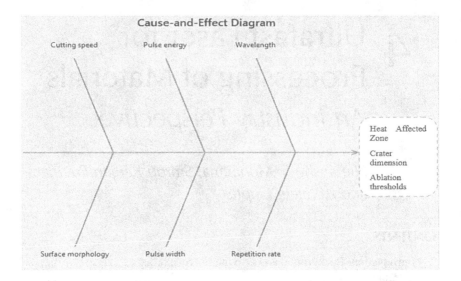

FIGURE 4.1   Cause-and-effect diagram of ultrafast laser processing.

or polymers with respect to the heat affected zone (HAZ), crater dimensions, laser fluencies, and ablation thresholds is seen in Figure 4.1.

4.2   ULTRAFAST LASER

Over the last 10–15 years, unrestricted free-form processing in 3D space on essential dimensions of 10 nm to 100 μm has been of the utmost interest. With a range of photo-polymers, captured in crystals, endless forms of objects were produced. This effort challenged the boundaries of newly discovered 3D manufacturing, and experimentation has been carried out with considerable throughput and resolution optimization (Matsumoto et al., 2013). The efforts carried out were with well-established standards built over a half-century in the fields of electronics in resolution, feature size, accuracy, and performance. The distinction between emerging 3D technology and established two-dimensional (2D) microtechnology is only partly due to the disparity between the character for 3D technology and the capability for processing laser techniques. Both methods have basic limits with respect to electrons and photons, respectively, set by fabrication wavelength. The difference in the corresponding wavelengths is in electron beam lithography. The de Broglie wavelength at a standard value of 50 kV acceleration is

$$\lambda e = \frac{h}{mv} = \sqrt{\frac{h^2}{2emV}} \approx 0.055 \ \text{Å} \qquad (4.1)$$

In which, e, m, and h are, respectively, the electron charge, mass, and Plank's constant, while in 3D-laser writing $\lambda_1 = 5000 \ \text{Å}$ (a green color).

After exploration of lasers, they now dominate the welding, cladding, and drilling on a special 3D robotic delivery capability at a scale of 0.1–10 m, where a stability point has been about 1 mm in the manufacturing of automotive for 40 years. Lasers with wavelengths of 1 μm or 10 μm and continuously in wave mode are effective, but they are not suitable for manufacturing tasks with a scale smaller than 1 mm. They are, thus, currently threatened by effective ultrashort pulsed lasers in accuracy and resolution. Double innovation was needed to create novel materials and laser resources that are best suitable to microprinting for the advancement of laser processing and not as a driving force in the long-praised manufacturing resolution.

Currently, the advent of a new and novel processing of effective fs lasers and support for nanoscale light-matter interaction science, manufacturing miniaturization progresses and innovating additive manufacturing, and surface texture have been in the cards to achieve the high-end use of modern technology.

4.3 DISRUPTIVE TECHNOLOGIES

The peculiarities of interaction in various forms at different positions of ultrashort distribution or within clear environments help describe protocols for processing material. Laser capability and fs pulse control sophistication meet the whole-time requirements of industries with respect to length, chirping, and polarization. Only time will tell the degree to which these approaches in real-world applications can become a scenario for this technology. Disruptive technologies are competing to pose accurate workpiece positioning and printing on a 3D scale. In the past, revolutionary, transformative innovations have proven commercially successful. A manufacturing process has been taken up with a special and appealing aspect of 3D printing by direct laser writing (DLW) in terms of industrial competitiveness.

4.4 INDUSTRIAL CHALLENGE

Manufacturing has its own rate of new investments being built and justified. The technology renovation part takes place a decade after the modernization cycle. The significant novel application for bringing an innovation to products is important and higher productivity is often needed. Lasers' procurement costs, efficiency, and durability have been often posed as the main driving factors for applications in industries using laser manufacturing. From the perspective of control, the interaction between light and matter are used, and in view of the functional necessity for high productivity, the breakthrough on an ultrashort pulsed laser is addressed.

Automotive, construction, and marking industries are already powered by cutting, welding, and additive manufacturing with CO_2, YAG, and fiber lasers working at 10 μm or 1 μm wavelengths. The speed and feed of operation are about 1 mm s^{-1} or higher by an average power of about 0.1–1 kW, which poses as a factor of realization. With turning points, it can be demonstrated in fs laser micromanufacturing to have a writing speed of 200 mm s^{-1} (Kato et al., 2005) on the screen of a cell phone. For energy transmission, a filament in water toward the propagation of beam has been another successful and restricted light delivery. This enables laser processing of curved surfaces in which close focusing involves axial location changes to match

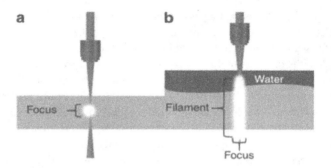

FIGURE 4.2 Beam focusing approaches.

the surface of the specimen precisely. Figure 4.2 depicts the principle of this method. The developed filament should have a length to width ratio of more than 2 mm in length where the diameter can be calculated experimentally (Rekstyte et al., 2013). For a fill exposure of volume on a 3D workpiece/pattern, the time of fabrication starts increasing quickly to a span of hours for 3D cross-sections of approximately 1 inch each.

Beam focusing approaches are performed for concentration of the beam on the surface or material bulk and for generation of a light filament in liquid medium for processing over various axial or surface curvature locations.

Mechanical galvo scanners use the quickest beam scan approach, which could deflect the beam much faster. For linear structures, a high-speed scanning is possible to achieve. Acceleration becomes a significant question when turning tight corners. The deflector based on an acousto-optical device allows increased scanning speeds even up to 2 m s^{-1} without apprehending for the turn radius, retaining high repeatability of positions. Also, sources of the laser can be made available for producing consecutive multibeams (Park et al., 2009) as well as using techniques for passive or active beam forming. These implementations significantly improve the throughput of fabrication. Tomographic data has recently been used as an approach for 3D printing based on computer-aided design (CAD) modeling.

This type of laser manufacturing is used where precision is necessary, such as to structure the surface and transparent material bulk that is quite hard and brittle. In case of layered materials and composites, they needed to be in a complex 3D fashion and can be carried out. The fundamentals will deliver energy with utmost accuracy where there are known effects of plasma reflection.

4.5 CONCLUSIONS AND FUTURE SCOPE

The sophistication of this 20-year-old industry is illustrated by certain reviews of the process. The fs laser manufacturing productivity needed for practical applications has been shown to be consistent on mature manufacturing techniques used in micromanufacturing, which is really a challenge. Biomedical applications are also an appropriate process with a physical review of the state-of-the-art technology. We conclude that we can solve this problem from a resolution point of view as well as efficiency of materials. Photo initiators are obsolete by controlled laser curing, and the interaction is controlled exclusively by certain parameters. This function is ventilated without additives and promoters of biopolymers. Proteins may be used in the future for applications organized to macroscale devices. In a nutshell, some future applications of ultrafast lasers can be applied to the integration of ultrafast laser-inscribed microstructures selectively etched with photonic components, monolithic integration of optical waveguides for near-infrared, selectively etched microfluidic channels/microcavities, and fabrication of complex embedded 3D structures with an ascent to optofluidic devices.

ACKNOWLEDGMENTS

The authors wish to thank NITTTR, Chandigarh, Hi-Tech Institute of Technology, Khurda.

REFERENCES

Davis K.M., Miura K., Sugimoto N., and Hirao K., 1996. Writing waveguides in glass with a femtosecond laser, *Optics Letters* 21, 1729–1731.

Glezer E.N., Milosavljevic M., Huang L., Finlay R.J., Her T.H. et al., 1996. Three-dimensional optical storage inside transparent materials, *Optics Letters* 21, 2023–2025.

Kato J.I., Takeyasu N., Adachi Y., Sun H.B., and Kawata S. 2005. Multiple-spot parallel processing for laser micronanofabrication, *Applied Physics Letters* 86, 044102.

Kawata S., Sun H.B., Tanaka T., and Takada K., 2001. Finer features for functional microdevices, *Nature* 412, 697–698.

Malinauskas M., Lukosevicius L., Butkus S., and Paipulas D. 2015. Femtosecond pulse light filament-assisted microfabrication of biodegradable polylactic acid (PLA) material, *Journal of Laser Micro Nanoengineering* 10, 222–228.

Malinauskas M., Rekstyte S., Lukosevicius L., Butkus S., Balciunas E. et al. 2014. 3D microporous scaffolds manufactured via combination of fused filament fabrication and direct laser writing ablation, *Micromachines* 5, 839–858.

Marshall G.D., Politi A., Matthews J.C.F., Dekker P., Ams M. et al. 2009. Laser written waveguide photonic quantum circuits. *Optics Express* 17, 12546–12554.

Maruo S., Nakamura O., and Kawata S. 1997. Three-dimensional microfabrication with two-photon-absorbed photopolymerization, *Optics Letters* 22, 132–134.

Matsumoto H., Unrath M., Zhang H.B., and Hainsey B. 2013. Laser direct ablation for patterning printed wiring boards using ultra-fast lasers and high speed beam delivery architectures, *Journal of Laser Micro Nanoengineering* 8, 315–320.

Mikutis M., Kudrius T., Slekys G., Paipulas D., and Juodkazis S. 2013. High 90% efficiency Bragg gratings formed in fused silica by femtosecond Gauss–Bessel laser beams, *Optical Materials Express* 3, 1862–1871.

Park S.H, Yang D.Y., and Lee K.S. 2009. Two-photon stereolithography for realizing ultraprecise three-dimensional nano/microdevices, *Laser & Photonics Reviews* 3, 1–11.

Rekstyte S., Malinauskas M., and Juodkazis S. 2013. Three-dimensional laser micro-sculpturing of silicone: towards bio-compatible scaffolds, *Optics Express* 21, 17028–17041.

Watanabe M., Sun H.B., Juodkazis S., Takahashi T., Matsuo S. et al. 1998. Three-dimensional optical data storage in vitreous silica, *Japanese Journal of Applied Physics* 37, L1527–L1530.

Section II

Material Characterization

5 Finite-Element Study of Influence of Build Orientation of Lattice Structures on Mechanical Properties

Prashant Kumar, Vijay Kumar Meena,
Tarun Panchal, Rahul Bhardwaj,
Rajesh Kumar, and Anjali Gupta

CONTENTS

5.1 INTRODUCTION

Many industries, like aerospace, automotive, and biomedical, are on a quest for stronger and lighter materials (Fleck et al., 2010; Vasconcellos et al., 2010; Hao et al., 2011; Yan et al., 2014; Bici et al., 2018; Burton et al., 2019). To fulfill this goal, various researchers have investigated the mechanical properties of lattice structures, which can be easily produced by additive manufacturing (AM) (Ashby, 2006). These materials

DOI: 10.1201/9781003203681-7

can also have better energy absorption and acoustic and thermal insulation character-istics (Hao et al., 2011). Ashby has stated that "a lattice structure should be treated as a material with its mechanical properties, allowing a comparison between the properties of a lattice structure and those of its parent material" (Ashby, 2006). Khaderi et al. (2014) have commented that lattice or cellular structures have more favorable specific characteristics (unit mass/volume properties) than their counterparts in bulk materials.

Lattice structures can be manipulated and manufactured in a variety of geometric structures with flexibility in the selection of shapes and materials using AM tech-niques, like direct metal laser sintering (DMLS), electron beam melting (EBM), extrusion free-forming (Challis et al., 2010; Wysocki et al., 2017), etc. Recent advancements in AM have allowed complex geometries to be produced with a rela-tively high degree of precision. Burton et al. (2019) have explored the versatility of design using AM to create porous lattices that increase the volume available for drug loading while preserving the load capacity of the hip implant. Vasconcellos et al. (2010) have studied in vivo influence of porous titanium implants on new bone ingrowth. Based on the results, it is concluded that increased porosity and pore size has a good effect on the amount of bone ingrowth. Hao et al. (2019) have designed an additively manufactured lightweight phase-change thermal controller structure for spacecraft based on lattice cells. The test results show that thermal efficiency is increased by 50% compared with traditional controllers with the same volume. Bici et al. (2018) have designed a wing panel section of aircraft using lattice structures to overcome the need for light and stiff structures in the aerospace industry.

Various mechanical characteristics of lattice structures depend on their geometric characteristics like unit cell type and size, the direction of loading, and porosity (Van Bael et al., 2012). Choy et al. (2017) have investigated the deformation and compressive prop-erties of Ti-6Al-4V lattice specimens manufactured by AM with the change in design, density, and orientation. The effectiveness of plateau stress for cubic lattice structure and compressive property values for honeycomb lattice structure has been significantly affected by the change of orientation. Weißmann et al. (2016) have determined material properties of the open porous scaffold design with a "twisted design" unit cell, which were manufactured by AM using Ti-6Al-4V. The unit cell orientations were changed and their mechanical properties were studied. Results have clearly shown the effect of orientation of the unit cells on Young's modulus, strain, and compressive strength.

The orientation of strut has a significant role in the assessment of mechanical characteristics for most lattice structures (Campoli et al., 2013). Therefore, mono-orientational mechanical behavior does not fully explain the elastic property behav-ior of anisotropic materials in real scenarios.

This study focuses on the mechanical effects of orientation of the unit cell of gyroid structure, which belongs to the triply periodic minimal surfaces (TPMS) fam-ily. TPMS is defined as "the surfaces with zero mean curvature and are specified by local area minimizing, which implies that any sufficiently small patch taken from the TPMS has the smallest area among all patches created under the same bound-aries" (Torquato and Donev, 2004). The DMLS three-dimensional (3D) printing process is used for manufacturing lattice structures. DMLS is a process generation of 3D complex parts through layer-by-layer manufacturing of powder layers under protective conditions.

The mechanical properties of gyroid lattice structures were investigated by comparing Young's modulus values at different orientations or phase shifts of unit cells (0°, 30°, 60°) in all three planes. The stress distribution, deformation behavior, and fatigue life are determined using finite-element analysis (FEA).

5.2 RESEARCH METHODS

5.2.1 DESIGN OF THE LATTICE STRUCTURES

Per ISO 13314 (titled Mechanical Testing of Metals–Ductility Testing–Compression Test for Porous and Cellular Metals), three cuboid porous specimens sized 10 × 10 × 20 mm is modeled using Simpleware ScanIP, version 2019.09 software. The relative volume fraction for the computer-aided design (CAD) lattice specimens has been kept to 35%. The gyroid unit cell dimension is 1.5 × 1.5 × 1.5 mm with a pore size of 0.90 mm for all models with different unit cell orientations. The three orientations of 0°, 30°, and 60° are selected and, accordingly, three different models of three different unit cell orientations are modeled (Figure 5.1). For the determination of porosity, a fully solid cuboid with the same dimensions as the test specimens has been modeled. For slicing and to generate support structures for 3D printing, Magics 24.0 software is used.

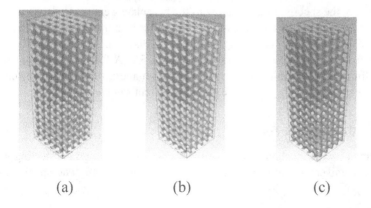

(a)   (b)   (c)

FIGURE 5.1 Designed model of gyroid lattice structure with unit cell orientations of (a) 0°, (b) 30°, and (c) 60°.

The spatial orientation of the unit cells has been rotated inside the outer rectangular form of the specimens for the characterization of material properties. The unit cells have been rotated by 0°, 30°, and 60° around all the three coordinate axes, respectively. The spatial position of nodes and the unit cells differed with each orientation, resulting in different configuration designs.

5.2.2 POROSITY CALCULATION

To verify the printability and the reproducibility of the fabrication process compared with CAD data, porosities for the specimens with different unit cell orientations were calculated. Porosities for the CAD models (Porosity$_{CAD}$) as well as for fabricated specimens (Porosity$_{DMLS}$) were calculated according to the following equations.

$$\text{Porosity}_{CAD} = \left(1 - V_{por}/V_{sol}\right) \times 100\%$$

where V_{por} is the volume of the CAD model of the porous structure and V_{sol} is the overall volume enclosed by the outer periphery.

$$\text{Porosity}_{DLMS} = \left(1 - W_{por}/W_{sol}\right) \times 100\%$$

where W_{por} is the weight of the manufactured porous specimen and W_{sol} is the weight of the manufactured solid cylinder of the same dimensions as the porous specimen. The weight is determined using scales.

5.2.3 MATERIAL USED AND MANUFACTURING

All the CAD models of the test specimens have been fabricated through DMLS using Ti-6Al-4V ELI powder provided by EOS and the EOS M290 metal printer. The samples were printed with an inert argon atmosphere for reducing the dangers of oxidation. The process parameters used for printing are tabulated in Table 5.1. After fabrication, the specimens attached with the baseplate were set for heat treatment in a muffle furnace for residual stress relieving. After heat treatment, the samples were removed from the baseplate using wire electric discharge machining. The samples parted from the baseplate are shown in Figure 5.2. A fully solid cuboid with the same dimensions as the test specimens was also manufactured and was used for the determination of weight difference among different samples.

5.2.4 COMPRESSION TESTING

All samples were mechanically tested using uniaxial compression testing with the help of a universal testing machine (UTM) (Figure 5.2(d)). The specimens were

TABLE 5.1
Sintering Process Parameters

Power	280 W
Type of laser	Nd-YAG laser
Diameter of laser beam	80 μm
Hatch distance	0.14 mm
Layer thickness	30 μm
Scan speed	1200 mm/s
Metal powder particle size	27–56 μm
Baseplate temperature	70°C

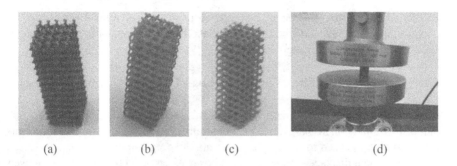

(a)   (b)   (c)   (d)

FIGURE 5.2 Specimens after heat treatment and cutting from the baseplate: lattice structures with unit cell orientations of (a) 0°, (b) 30°, and (c) 60°. (d) Tested sample between the compression anvils of the UTM.

tested with three samples per unit cell orientation to reduce the variability of results. Compressive testing of the specimens was performed at a fixed strain rate of 0.005 mm/min. All tests were done under normal standardized conditions. The values of Young's modulus (E) have been deduced by the test results.

5.2.5 Finite-Element Analysis (FEA)

FEA was used to determine the stress distribution and deformation at applied force F of 1000 N for the lattice design samples with different unit cell orientations. For FEA, the meshing of lattice structures is done using Simpleware ScanIP 2019.09 with the +FE free meshing algorithm and tetrahedra (linear) element type (Figure 5.3(a)). The file is then transferred to Abaqus CAE (commercial software) for FEA. Young's modulus of the specimens derived from the uniaxial compression test was utilized

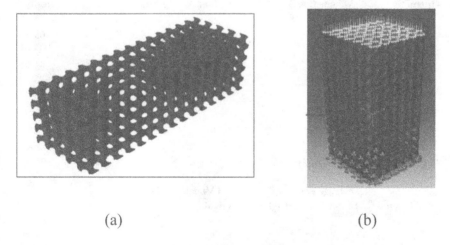

(a)   (b)

FIGURE 5.3 (a) Mesh representation of Schoen gyroid lattice structure and (b) Loading and boundary conditions visualized.

for FEA. Fatigue strength was also simulated using the software fe-safe at cyclic loads of ±1000 N, respectively, for 106 cycles. The minimum fatigue life cycle is calculated using the following formula:

$$\text{Fatigue Life Cycle} = 10 \wedge \big((\text{Log} - \text{Life Repeats})\big)/2$$

The uppermost surface of the lattice structures is loaded with the compressive force and the lowermost surface is given fixed constraints (Figure 5.3(b)). Quasi-static simulation analysis is used to mimic the real static loading conditions. The von Mises stress results and displacement results are chosen as output. These results were used for the comparative study of the gyroid lattice structures of three different types of orientations.

5.3   RESULTS

5.3.1   PorosityResult

The volume of CAD models of lattice structures and the solid model is found from the software and is used to calculate the CAD porosity. For the fabricated part, the weight is taken and tabulated. Tables 5.2 and 5.3 summarized the results.

The examination of the developed models shows small differences between the porosity of CAD and fabricated models (below 2%).

TABLE 5.2
CAD Model Porosity

Orientation	Lattice Structure Volume (mm³)	Solid Model Volume (mm³)	Relative Volume Fraction (%)	Porosity (%)
0°	689.42	2000.00	34.47	65.53
30°	687.87	2000.00	34.39	65.61
60°	690.07	2000.00	34.50	65.50

TABLE 5.3
Fabricated Model Porosity

Orientation	Average Weight of Lattice Structure (g)	Solid Model Weight (g)	Relative Volume Fraction (%)	Porosity (%)
0°	3.19	8.82	36.17	63.83
30°	3.17	8.82	35.98	64.02
60°	3.21	8.82	36.36	63.64

5.3.2 MECHANICAL PROPERTIES

All manufactured samples were compression tested until failure took place. The results for Young's modulus are listed in Table 5.4. In Figure 5.4 representative samples are illustrated for each design type and the site of fracture is depicted. All the samples behave similarly as their representative samples for each unit cell orientation.

The highest modulus was found for the unit cell orientation of 60° with a Young's modulus of approximately 1513.85 MPa. The lowest modulus was found for the unit cell orientation of 30° with a Young's modulus of approximately 1351.21 MPa.

TABLE 5.4
Fabricated Model Porosity

Orientation	Young's Modulus (MPa)
0°	1387.217
30°	1351.210
60°	1513.849

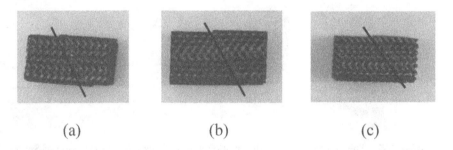

| (a) | (b) | (c) |

FIGURE 5.4   Overview of tested samples of unit cell orientations of (a) 0°, (b) 30°, and (c) 60°. Fracture lines are identified by a black line.

5.3.3 STRESS DISTRIBUTION AND DEFORMATION RESULTS OF FEA

The FEA results for maximum von Mises stress and maximum displacement for a compressive load of 1000 N are listed in the following in Table 5.5 and shown in Figure 5.5 for all the design variations.

The highest stress was found for the unit cell orientation of 60° with maximum von Mises stress of approximately 332.10 MPa. The lowest stress was found for the unit cell orientation of 30° with maximum von Mises stress of approximately 228.50 MPa. The highest deformation was found for the unit cell orientation of 30° with a maximum displacement value of approximately 1.91 mm. The lowest modulus was found for the unit cell orientation of 60° with a maximum displacement value of approximately 1.59 mm.

TABLE 5.5
Maximum von Mises Stress and Maximum
Displacement for Design Variations

Orientation	Maximum von Mises Stress (MPa)	Maximum Displacement (mm)
0°	248.00	1.654
30°	228.50	1.914
60°	332.10	1.596

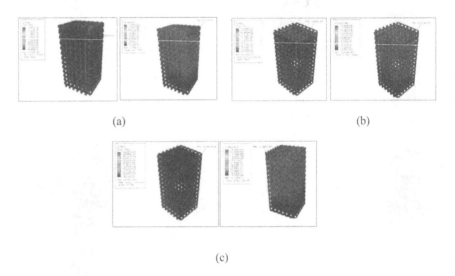

(a)   (b)

(c)

FIGURE 5.5 The von Mises stress distribution and deformation representation of Schoen gyroid lattice with unit cell orientations of (a) 0°, (b) 30°, and (c) 60°.

5.3.4 FATIGUE STRENGTH RESULTS FROM FEA

The minimum log-life repeats and minimum fatigue cycles up to which no crack can initiate at a cyclic load of 1000 N for the models are listed in Table 5.6 and representative diagrams are shown in Figure 5.6.

5.4 DISCUSSION

Various studies often suggest that the build orientation has a significant role in the mechanical properties of AM full and open porous structures. In this study, the effect of the unit cell orientation of Schoen gyroid lattice structures on their mechanical properties was studied with the help of FEA.

TABLE 5.6
Minimum Log-Life Repeats Value and
Minimum Fatigue Life for a Cyclic Load of
1000 N for Lattice Structures

Orientation	Minimum Log-Life Repeats	Minimum Fatigue Life in Cycles
0°	6.207	8.053×10^5
30°	6.081	6.025×10^5
60°	6.613	2.051×10^6

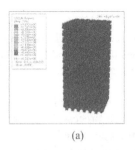 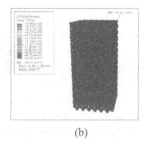 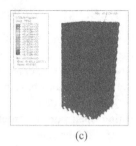

(a) (b) (c)

FIGURE 5.6 Log-life repeats result in the representation deduced by FEA for unit cell orientations of (a) 0°, (b) 30°, and (c) 60°.

Comparisons of the calculated porosities for the fabricated models result in high accordance with the porosities derived from CAD data with deviations less than 2%. This shows the excellent manufacturing quality of the samples.

The unit cell orientation of 60° exhibits better mechanical properties from the other two orientations, as it has greater laser scan areas than the other two orientations as shown in Figure 5.7. As the area scanned by the laser is more for the 60° orientation than the other two orientations, the elastic modulus is greater for this orientation and has less deformation as verified by FEA results.

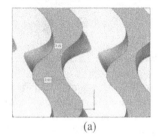 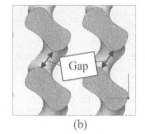 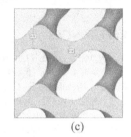

(a) (b) (c)

FIGURE 5.7 Showing the maximum and minimum length of cross-section scanned by laser for unit cell orientations of (a) 0°, (b) 30°, and (c) 60°.

The orientation of the unit cell thus has a crucial influence on the strength and deformation behavior. This study shows that the orientation of the unit cell can influence the mechanical properties of the open lattice structure. While using an open lattice structure for any application, the unit cell orientation can affect the buildability as well as load requirement for that application. By orientating the unit cell, the structure growth direction can be changed from against the recoater blade toward the direction of the recoater and, accordingly, the orientation angle can be selected for the required mechanical properties.

The different defined areas can fulfill their function and mechanical requirement only by varying the orientation of a unit cell. It can also help in increasing the printability of the model by changing the orientation, and structures can be rotated to grow in the direction of the recoater blade in DMLS.

5.5   CONCLUSIONS

The Young's modulus of Ti-6Al-4V gyroid lattice samples fabricated via DMLS is experimentally examined at different unit cell orientations (0°, 30°, and 60°). The stress distribution, deformation results, and fatigue life have been deduced by FEA. Some conclusions have been described as follows:

It has been demonstrated that high geometrical accuracy is achievable for fabricating open porous structures of Ti-6Al-4V using the DMLS process. The porosity of the fabricated structures differed less than 2% from the idealized porosity of the structures as calculated by CAD. Based on compression testing under uniaxial loading conditions, the influence of cell orientation on Young's modulus was listed. Based on the FEA results, influence of cell orientation on the induced stress and deformation was demonstrated. The orientation of the unit cell results in a change of its mechanical properties, which can be due to (1) the scan area and the material deposited on a given plane and (2) the position of the structures toward the recoater force. The influence on the unit cell orientation on the mechanical properties of lattice structures makes it necessary to consider it when planning and designing the lattice structures for any given loading condition and working environment.

ACKNOWLEDGMENT

This work was supported by the SERB, New Delhi, India (Grant EEQ/2017/000154).

REFERENCES

Ashby M.F., 2006. The properties of foams and lattices, *Philosophical Transactions. Series A* 364(1838), 15–30.
Bici M., Brischetto S., Campana F., Ferro C. G., Seclì C., Varetti S., Maggiore P., and Mazza A., 2018. Development of a multifunctional panel for aerospace use through SLM additive manufacturing, *Procedia CIRP* 67, 215–220.
Burton H.E., Eisenstein N.M., Lawless B.M., Jamshidi P., Segarra M.A., Addison O., et al., 2019. The design of additively manufactured lattices to increase the functionality of medical implants, *Materials Science and Engineering* 94, 901–908.

Campoli G., Borleffs M.S., Amin Yavari S., Wauthle R., Weinans H., and Zadpoor A.A., 2013. Mechanical 4. properties of open-cell metallic biomaterials manufactured using additive manufacturing, *Materials & Design* 49, 957–965. doi: https://doi.org/10.1016/j.matdes.2013.01.071

Challis V.J., Roberts A.P., Grotowski J.F., Zhang L.C., and Sercombe T., 2010. Prototypes for bone implant scaffolds designed via topology optimization and manufactured by solid freeform fabrication, *Advanced Engineering Materials* 12(11), 1106–1110.

Choy S.Y., Sun C.N., Leong K.F., and Wei J., 2017. Compressive properties of Ti-6Al-4V lattice structures fabricated by selective laser melting: design, orientation and density, *Additive Manufacturing* 16, 213–224.

Fleck N.A., Deshpande V., and Ashby M.F., 2010. Micro-architectured materials: past, present and future, *Proceedings of the Royal Society A* 466(2121), 2495–2516.

Hao L., Raymont D., Yan C., Hussein A., and Young P., 2011. Design and additive manufacturing of cellular lattice structures, Proceedings of International Conference on Advanced Research in Virtual and Rapid Prototyping (VRAP), Leiria, Portugal.

Hao Z., Zhang X., Huizhong Z., Huning Y., Hongshuai L., Xiao L., and Yaobing W., 2019. Lightweight structure of a phase-change thermal controller based on lattice cells manufactured by SLM, *Chinese Journal of Aeronautics* 32(7), 1727–1732.

Khaderi S., Deshpande V., and Fleck N., 2014. The stiffness and strength of the gyroid lattice, *International Journal of Solids and Structures* 51(23–24), 3866–3877.

Torquato S., and Donev A., 2004. Minimal surfaces and multifunctionality, *Proceedings of the Royal Society London A* 460(2047), 1849–1856.

Van Bael S., Chai Y.C., Truscello S., Moesen M., Kerckhofs G., Van Oosterwyck H., and Schrooten J., 2012. The effect of pore geometry on the in vitro biological behavior of human periosteum-derived cells seeded on selective laser-melted Ti6Al4V bone scaffolds. *Acta Biomaterialia* 8(7), 2824–2834. doi: https://doi.org/10.1016/j.actbio.2012.04.001

Vasconcellos L.M.R.D., Leite D.O., Oliveira F.N.D., Carvalho Y.R., and Cairo C.A.A., 2010. Evaluation of bone ingrowth into porous titanium implant: histomorphometric analysis in rabbits, *Brazilian Oral Research* 24(4), 399–405.

Weißmann V., Bader R., Hansmann H., and Laufer N., 2016. Influence of the structural orientation on the mechanical properties of selective laser melted Ti6Al4V open-porous scaffolds, *Materials & Design* 95, 188–197.

Wysocki B., Maj P., Krawczyńska A., Rożniatowski K., Zdunek J., Kurzydłowski K. J., and Święszkowski W., 2017. Microstructure and mechanical properties investigation of CP titanium processed by selective laser melting (SLM), *Journal of Materials Processing Technology* 241, 13–23. doi: https://doi.org/10.1016/j.jmatprotec.2016.10.022

Yan C., Hao L., Hussein A., Bubb S.L., Young P., and Raymont D., 2014. Evaluation of lightweight AlSi10Mg periodic cellular lattice structures fabricated via direct metal laser sintering, *Journal of Materials Processing Technology* 214(4), 856–864.

6 Surface Protection and Restoration of Mild Steel under Coating Technology
Cold Spray Additive Manufacturing

Abdul Faheem, Faisal Hasan,
Mohd. Gulam Waris Khan,
and Qasim Murtaza

CONTENTS

6.1 INTRODUCTION: BACKGROUND AND DRIVING FORCES

Corrosion is the deterioration or disintegration of any material into an unstable state by reacting with the environment. It is highly dangerous for engineering structures (i.e., bridges, pipelines, naval ships, aerospace industries, buildings, different plants, etc.). Furthermore, it oxidizes the surface of the metal or alloys consequently damaging the entire surface of the objects. This severe problem not only deteriorates the materials but also adds an extra burden of billions of dollars to the economy of any country. Apart from that, when any material becomes corroded, it loses its dimensional stability. It causes the parts/object to replace the dimensional accurate part, which also puts an extra burden on a country's economy. Cold spray (CS) or cold gas dynamic spray (CGDS) or solid-state particle deposition (SSPD) is an additive manufacturing (AM) approach used to repair metallic/nonmetallic structures (Assadi et al., 2003; Bae et al., 2008; Cavaliere and Silvello, 2014; Zhou et al., 2011). This AM approach is to repair and stabilize the dimension by using a high velocity of microparticles/nanoparticles with the carrier gas using

DOI: 10.1201/9781003203681-8

the supersonic nozzle. This technology potentially controls the corrosion damage related to metallic, alloy, and ceramic structures. Furthermore, different surfaces of different types of equipment in public places, especially in railways, hospitals, shopping malls, etc., which cause bacterial contamination and a serious threat should be reduced by proper selection of materials. Different studies have been examined for controlling the corrosion rate, corrosion resistance, different particle/substrate coating, hard/soft coating, etc. (Yildirim et al., 2011; Murtaza et al., 2012; Moridi et al., 2014; Tan et al., 2017; Faheem et al., 2019; Yeom et al., 2019,). Balani et al. (2005) analyzed the electrochemical characteristics of CS AA 1100 onto the same substrate using helium gas. From an economic perspective, the cathodic coating has been opted for different metals/alloys/composites to improve the corrosion resistance, but the corrosion resistance of CS metal coatings also gave better results. Composite coating through AM also provides a drastic enhancement in the results, especially in porosity, cracks, etc. Da Silva et al. (2018) studied the corrosion behavior of WC/ Co coating onto AA 7075-T6 through CS and found a potential coating of carbide. In this study, enhancement of corrosion resistance and repair of deteriorating parts have been done by using micro-sized particles accelerated with high velocity onto the substrate using the AM CS approach

6.2   EXPERIMENTAL AND NUMERICAL APPROACH

The experimental approach has been done by taking the particles in the range of 5–20 μm. In this study, the de Laval convergent-divergent nozzle has been used to achieve the metallic coating and enhancement of corrosion resistance. The CGDS setup using the AM approach as shown in Figure 6.1 consists of a nozzle, powder preheating device, pressure sensor, gas preheater, etc. The carrier gas used in this study is nitrogen because it has good capability to cause the microparticle to achieve high velocity, as in the previous studies. The velocity of the particle is in the range of approximately 600 m/s with the preheated temperature of the particle and substrate (mild steel). Furthermore, a numerical approach (finite-element methodology [FEM]) using Abaqus/Explicit has been used to understand the coating phenomenon

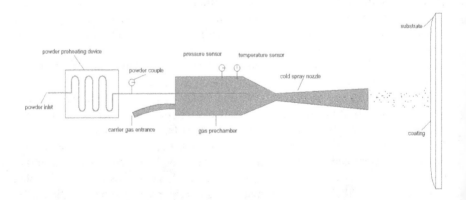

FIGURE 6.1   The setup of cold spray.

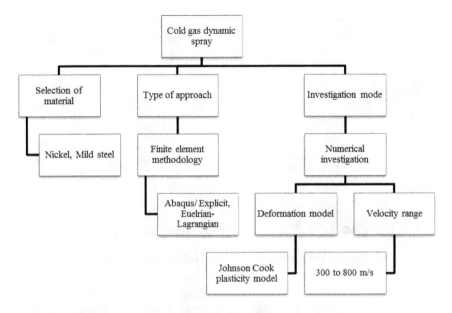

FIGURE 6.2   Flowchart of cold spray under numerical modeling.

on the substrate. The plasticity model for the deformation of particle and substrate along with the damage criteria is the Cook plasticity model. In Figure 6.2, a flowchart of numerical modeling using CS shows the details of the investigation.

6.3   RESULTS AND DISCUSSION

The enhancement of corrosion resistance under kinetic spray has been investigated using the AM approach. In Figure 6.3, transmission electron microscopy (TEM) for the characterization of the metallic coating and corrosion resistance approach is used. Scanned electron microscopy (SEM) for the characterization of the metallic

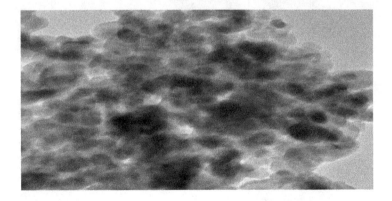

FIGURE 6.3   Transmission electron microscopy (TEM) of nickel powder.

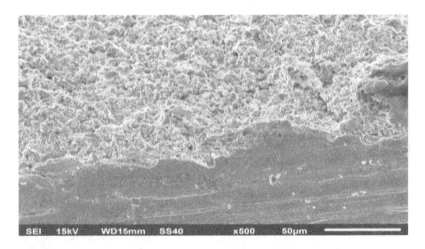

FIGURE 6.4 Scanned electron microscopy (SEM) of the nickel coating.

coating is shown in Figure 6.4. SEM clearly shows plastic deformation of the micro-sized particle and formation of the metallic coating. The multilayer coating has been created so that resistance of corrosion for the substrate is increased. In this study, some particles deformed plastically more compared with the other particle.

FEM provides suitable results under different criteria of analysis. Under high velocity (500 m/s), the deformation pattern, stress, and pressure of nickel have been studied (Figure 6.5). This deformation behavior is very helpful in understanding how to restore the corroded surface and gain the dimension stability of the substrate.

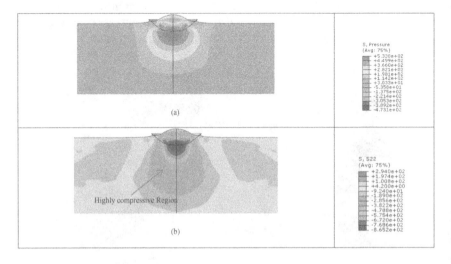

FIGURE 6.5 Stress distribution: (a) pressure stress and (b) normal stress.

6.4 CONCLUSIONS

In this investigation, the CSAM approach has been used to restore the dimension accuracy of the corroded material. Apart from the experimental approach, numerical investigation helps us to create a metallic coating on the substrate and deformation pattern of the particle. The characterization technique called SEM pillared the coating on the substrate. Moreover, mechanical properties, corrosion resistance, and wear resistance on the surface were gained. Meanwhile, the numerical approach showed the deformation and deposition of micro-sized particles on the substrate.

REFERENCES

Assadi H., Ga F., Stoltenhoff T., and Kreye H., 2003. Bonding mechanism in cold gas spraying, *Acta Materialia* 51, 4379–4394.

Bae G., Xiong Y., Kumar S., Kang K., and Lee C., 2008. General aspects of interface bonding in kinetic sprayed coatings, *Acta Materialia* 56, 4858–4868.

Balani K., Laha T., Agarwal A., Karthikeyan J., and Munroe N., 2005. Effect of carrier gases on microstructural and electrochemical behavior of cold-sprayed 1100 aluminum coating, *Surface and Coatings Technology* 195, 272–279. doi: https://doi.org/10.1016/j.surfcoat.2004.06.028

Cavaliere P., and Silvello P., 2014. Processing parameters affecting cold spay coatings performances, *International Journal of Advanced Manufacturing Technology*, 71, 263–277.

Da Silva F.S., Cinca N., Dosta S., Cano I.G., Couto M., Guilemany J.M., and Benedetti A.V., 2018. Corrosion behavior of WC-Co coatings deposited by cold gas spray onto AA 7075-T6, *Corrosion Science* 136, 231–243. doi: https://doi.org/10.1016/j.corsci.2018.03.010

Faheem A., Hasan F., and Murtaza Q., 2019. Material perspective and deformation pattern of micro-sized metallic particle using cold gas dynamic spray. *Advances in Computational Methods in Manufacturing. Lecture Notes on Multidisciplinary Industrial Engineering.* Springer, Singapore. doi: https://doi.org/10.1007/978-981-32-9072-3_34

Moridi A.M., Azadi M., and Farrahi G.H., 2014. Surface & coatings technology thermo-mechanical stress analysis of thermal barrier coating system considering thickness and roughness effects, *Surface Coating Technology* 243, 91–99.

Murtaza Q., Stokes J., and Ardhaoui J., 2012. Experimental analysis of spray dryer used in hydroxyapatite thermal spray powder, *Journal of Thermal Spray Technology* 21, 963–974.

Tan, A.W.Y., Sun W., Phang Y.P., Dai M.H., Marinescu I., Dong Z.L., and Liu E.J., 2017. Effects of traverse scanning speed of spray nozzle on the microstructure and mechanical properties of cold-sprayed Ti6Al4V coatings, *Journal of Thermal Spray Technology* 26, 1484–1497.

Yeom H., Dabney T., Johnson G., Maier B., Lenling M., and Sridharan K., 2019. Improving deposition efficiency in cold spraying chromium coatings by powder annealing, *International Journal of Advanced Manufacturing Technology* 100, 1373–1382.

Yildirim B., Muftu S., and Gouldstone A., 2011. Modeling of high velocity impact of spherical particles, *Wear* 270(9), 703–713.

Zhou X., Wu X., Wang J., and Zhang J., 2011. Numerical investigation of the rebounding and the deposition behavior of particles during cold spraying, *Acta Materialia Sin* 24(1), 45–53.

7 Experimental Investigation and Optimization of EDM Parameters During Machining of Al/B$_4$C/Gr MMC

Mumtaz Rizwee, P. Sudhakar Rao,
and Mohd Yunus Khan

CONTENTS

7.1 INTRODUCTION

Metal matrix composite (MMC) materials have become the most attractive materials for industry among all novel developed materials because of their remarkable mechanical properties such as toughness, hardness, wear and corrosive resistance, high temperature resistance, etc. Yet, because of abrasiveness, hardness, and brittleness, it is very difficult to machine such material by a nonconventional machining process. Because of high tool wear rate and high tooling cost during machining of

DOI: 10.1201/9781003203681-9

such material through the conventional machining process, a noncontact machining process such as electro discharge machining (EDM) is more suitable for machining such material (Lau et al., 1995; Singh et al., 2004; Khan and Rao, 2019). Many researchers have investigated the EDM of ceramic and MMC material (König et al., 1988; Rizwee and Rao, 2018; Rizwee et al., 2019). Performance of EDM is evaluated in terms of material removal rate (MRR), TWR, EWR, SR, ROC, etc. (Rizwee et al., 2021; Rizwee and Rao, 2021). Fattouh et al. (1990) established the mathematical relation by using response surface modeling (RSM) among the process parameter and process responses during the EDM process (Rao et al., 2016). Khan et al. (2020a–c) investigated the effect of biodiesel and powder mixed dielectric fluid in the EDM process (Rao et al., 2014).

7.2   WORKPIECE MATERIAL AND REINFORCEMENT

Boron carbide (B_4C) with less than a 10-μm diameter that is 99% pure and graphite (Gr) with less than 20-μm reinforcement were purchased from a local vendor. Aluminum alloy 7075 (Al-7075) in the form of cylindrical ingot with a diameter of 40 mm and length of 380 mm was purchased from Bharat Aerospace Limited, Mumbai, India. The chemical composition of Al-7075 alloy is given as Al 89.60%, Zn 5.70%, Mg 2.30%, Cu 1.50, Fe 0.41%, Cr 0.18%, Si 0.13%, Mn 0.13%, and Ti 0.05%.

7.3   EXPERIMENTAL DETAILS

The experimental setup was developed in the Indo Danish Tool Room (IDTR) Jamshedpur, Jharkhand, India, to conduct the present research work. In this work, Al-7075 alloy was stir-casted with different combinations of reinforcement 3wt.% B_4C and 7% Gr, 5% B_4C and 5% Gr, 7% B_4C and 3% Gr, respectively, were machined using a copper electrode with an 18–mm diameter on the OSCARMAX EDM machine manufactured by OSCARMAX EDM Ltd., Taichung, Taiwan (Figure 7.1),

FIGURE 7.1   OSCARMAX electro discharge machine.

TABLE 7.1
Input Parameter with Their Levels

Input Parameter	Low Level (−1)	Medium Level (0)	High Level (+1)
Servo voltage (V)	3	6	9
Spark on time (μs)	300	400	500
Pulse current (A)	2	4	6
Sample	1	2	3

in the IDTR. Negative polarity was maintained for the tool and positive polarity for the workpiece during the experimental work. EDM fluid SE 180 was used as a dielectric fluid to perform the experiment. In the present investigation, four independent parameters, such as servo voltage (3, 6, and 9 V), spark on time (300, 400, and 500 μsec), pulse current (2, 4, and 6 amps), and different samples (samples 1, 2, and 3) and their different levels were fixed (Table 7.1). Remaining input parameters like spark off time were kept constant at 150 μsec and the tool workpiece gap was kept constant at 10 mm. Different levels of the independent variable were selected through studying the literature survey (Dhar et al., 2007; Senthilkumar and Omprakash, 2011). Based on the Taguchi methodology, a total of nine experiments were planned (Table 7.2). Five different MRR responses were studied during the experimental work.

TABLE 7.2
Experimental Matrix Using L9 Orthogonal Array

Experiment No.	Servo Voltage (V)	Spark on Time (μs)	Pulse Current (A)	Sample
1	3	300	2	1
2	3	400	4	2
3	3	500	6	3
4	6	300	4	3
5	6	400	6	1
6	6	500	2	2
7	9	300	6	2
8	9	400	2	3
9	9	500	4	1

7.4   MEASUREMENT OF RESPONSE VARIABLE

MRR is expressed as the ratio of the difference of the weight of the workpiece before machining and after machining to the time of machining. Mathematically it can be expressed as

$$\text{MRR} = \frac{W_{jb} - W_{ja}}{t}$$

TABLE 7.3

Experimental Finding for Process Responses

Experiment No.	Servo Voltage (V)	Spark on Time (μs)	Pulse Current (A)	Sample	MRR (g/min)
1	3	300	2	1	0.004759
2	3	400	4	2	0.021296
3	3	500	6	3	0.062490
4	6	300	4	3	0.024560
5	6	400	6	1	0.075189
6	6	500	2	2	0.002949
7	9	300	6	2	0.077029
8	9	400	2	3	0.002821
9	9	500	4	1	0.026167

where

W_{jb} = weight of the work piece before machining

W_{ja} = weight of work piece after machining

t = time of machining

Weight of the workpiece before and after machining was measured by digital balance, which was manufactured by Testing Instruments Mfg. Co. Pvt. Ltd. The list count of the digital weight machine was 0.1 mg. The experimental findings for MRR have been tabulated in Table 7.3.

7.5 DEVELOPMENT OF MATHEMATICAL MODELING

7.5.1 REGRESSION ANALYSIS

The regression mathematical equation commonly used is represented as

$$Y = f \left(\text{Servo voltage (SVO), spark on time} \left(T_{ON} \right), \text{pulse current (I), sample (S)} \right)$$

(7.1)

Y is the process response, such as MRR; f is the response function; and SVO, T_{ON}, I, and S represent the process variables. The fitness of the experimental finding was demonstrated through the simple regression analysis methodology. MINITAB statistics was used to perform the regression analysis. A first-order nonlinear polynomial model was used to develop the relation between process responses and process variable. A first-order nonlinear polynomial model is shown below.

$$Y = \beta_0 + \beta_1 A + \beta_2 B + \beta_3 C + \beta_4 D + \varepsilon$$

(7.2)

where β_1, β_2, β_3, and β_4 are the input parameters and ε is the error. An empirical equation was expressed to develop the functional relationship between process variables and process responses.

The regression equation for MRR is as follows:

$$\text{MRR} = -0.0256 + 0.00097\,\text{SVO} - 0.000025\,T_{ON} + 0.01701\,\text{I} - 0.00271\,\text{S}\ldots\left(R^2\,94.97\right)$$

$$(7.3)$$

7.6 RESULTS AND DISCUSSION

7.6.1 ANALYSIS OF HARDNESS

A prepared sample after polishing was subjected to the hardness test. An Optical Brinell Hardness Tester (manufactured by Fine Spavy Associates and Engineering Pvt. Ltd.) Model OPAB-3000 N was used for the hardness test. A tungsten ball with 5-mm diameter and 750-kg load was used to perform the test. Hardness tests were conducted on the composite materials and the results are listed in Table 7.4. Hardness increases when increasing the percentage of B_4C and decreases on increasing the percentage of graphite. Because B_4C is abrasive and brittle in nature, hardness enhances when increasing its percentage, but Gr is brittle and soft in nature, which reduces the hardness when increasing its percentage.

TABLE 7.4
Hardness Value of Stir Casted AMMC with B_4C and Graphite Reinforcement

Sample	Observed Values			
	Hardness (BHN)			
	I	II	III	Average
Sample 1 (B_4C 3% and Gr 7%)	127	129	130	129
Sample 2 (B_4C 5% and Gr 5%)	131	134	133	133
Sample 3 (B_4C 7% and Gr 3%)	140	137	137	138

7.6.2 MICROSTRUCTURE ANALYSIS

Microstructure analysis helps to see the microstructure of the materials. It helps to understand which behavior changes in the microstructure take place during different processes, such as the casting process, heat treatment process, machining process, etc. A mirror-like smooth surface is obtained through polishing the cast $Al/B_4C/Gr$ composites because at higher magnification it will help to expose the grain boundary and secondary phases visible in the microscope as shown in Figure 7.2. The microstructure and interface characteristic between reinforcements and matrices affects the property of composites materials. Figure 7.2 displays the optical microstructures of different percentages of B_4C- and Gr-reinforced aluminum metal matrix composites (AMMCs). Microstructure analysis revealed that there was no indication of agglomeration of reinforcements in the MMCs because of the nonvariation of contact

Al7075+7% B₄C+3% Gr   Al7075+5% B₄C+5% Gr   Al7075+3% B₄C+7% Gr

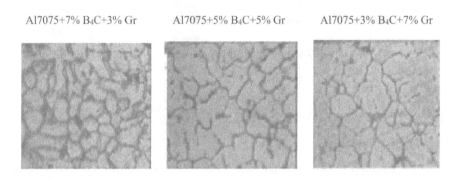

FIGURE 7.2   Microstructure view of Al/B₄C/Gr MMC materials.

time between B₄C and Gr particles in the liquid Al during composite formation and uniform distribution of the reinforcement's particles during the stirring process.

7.6.3 STUDY OF S/N RATIO

The S/N ratio is defined as the signal-to-noise ratio. Signal (S) is referred to as the desired value and noise (N) is referred as the undesired value. Taguchi analysis uses the S/N ratio to determine the quality characteristics. The S/N ratio is mathematically represented as

$$\text{S/N ratio for "Larger is Better" } SN_L = -10\log\left(\frac{1}{n}\sum_{i=1}^{n}\frac{1}{y_i^2}\right) \text{ for } i = 1 \text{ to } n \quad (7.4)$$

$$\text{S/N ratio for "Smaller is Better" } SN_S = -10\log\left(\frac{1}{n}\sum_{i=1}^{n}y_i^2\right) \text{ for } i = 1 \text{ to } n \quad (7.5)$$

where y_i is the value of MRR for the ith test and n is the total number of tests. To improve the machining rate MRR should be high. Hence, larger and better quality characteristics were selected for the MRR. The S/N ratio value was calculated by using the Eq. (7.4). The value of MRR was calculated from the experimental work and the S/N ratio value is tabulated in Table 7.5. An optimum process parameter was selected from S/N ratio calculation data.

TABLE 7.5
S/N Ratio for MRR

Level	SVO	T_{ON}	I	S
1	−34.66	−33.64	−49.35	−33.52
2	−35.09	−35.63	−32.42	−35.44
3	−34.97	−35.45	−22.94	−35.76

7.6.4   EDM PARAMETRIC EFFECT ON MRR

MRR is one of the important factors of machining that helps to increase the productivity and reduce the production time of any product. Equation (7.3) was developed through experimental findings and regression analysis, which helps to know the effect of the input process parameter on the MRR. The value of R^2 and adjusted R^2 are 94.97% and $R^2 = 89.93\%$, which indicates that the regression model fitted well between the independent variable (factor) and dependent variable (MRR).

Figure 7.3 indicates that when increasing the T_{ON}, MRR starts to decrease slightly. T_{ON} means the time during which the spark is generated. The enhancement of T_{ON} indicates that the same heating effect is applied for a longer period of time. This will enhance the amount of heat that is conducted into the workpiece, which will increase the MRR (Natsu et al., 2006; Panda, 2008). But on increasing more discharge duration, the pressure inside the plasma channel will become lower and MRR will start to decrease (Eubank et al., 1993). It is clearly observed in the main plot of MRR (Figure 7.3) and ANOVA (Table 7.6) that the pulse current (I) has a major effect on MRR. MRR increases in nonlinear fashion of increasing the I. The higher the pulse current means the workpiece is subjected to higher thermal energy, which helps to enhance the crater size; as a result, MRR increases (Müller and Monaghan, 2000; Rao et al., 2015). Further, on increasing the percentage of B_4C and decreasing the percentage of GR, MRR decreases slightly. B_4C and Gr both are ceramic particles and have very high melting points like SiC particles; hence, both are not melted during the EDM process (Hung et al., 1994; Müller and Monaghan, 2001; Rao et al., 2017). In this case, removal of B_4C and Gr particles from the AMMC takes place through melting and vaporization of the surrounding aluminum metal matrix of reinforced

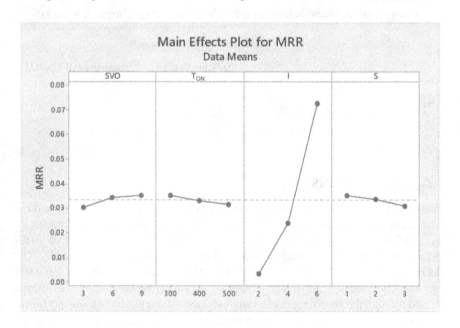

FIGURE 7.3   Effect of EDM parameters on MRR.

TABLE 7.6
ANOVA for MRR

Source	df	Seq. SS	Adj. SS	Adj. MS	% Contribution
SVO	2	0.000057	0.000057	0.000029	0.76
T_{ON}	2	0.000136	0.000136	0.000068	1.82
I	2	0.007214	0.007214	0.003607	96.78
S	2	0.000046	0.000046	0.000023	0.62
Total	8	0.007454			100

particles up to an extent where B_4C and Gr particles become detached. Hence, MRR decreases when increasing the B_4C particle. At the same time, when increasing the Gr percentage, MRR also increases because Gr has more thermal conductivity compared with B_4C, which enhances the melting and evaporation mechanism during the EDM process. Hence, MRR increases when increasing the Gr percentage. SVO slightly affected the MRR. When increasing the servo voltage, MRR increases slightly as shown in Figure 7.3.

Table 7.7 shows the comparative analysis between predicted and experimental results for MRR. It is seen that predicted and experimental values for MRR are probably same.

TABLE 7.7
Comparative Analysis Between Predicted and Experimental Results

		Optimum Machining Parameter	
		Predicted Result	**Experimental Result**
MRR	Level	A1B1C3D1	A1B1C3D1
	MRR	0.0728187	0.0786518
	S/N ratio for MRR	−20.0428	−22.0858

7.7 CONCLUSIONS

Present experimental analysis highlights that MMR is greatly affected by the dominant input parameter during EDM of the $Al/B_4C/Gr$ MMC. When increasing the T_{ON}, MRR starts to decrease slightly. Pulse current has a major effect on MRR. When increasing the percentage of B_4C and decreasing the percentage of Gr, MRR decreases slightly. SVO slightly affected the MRR. When increasing the servo voltage MRR increases slightly. The fitness of the experimental finding was demonstrated through the simple regression analysis methodology. ANOVA was performed to investigate the effect of the process parameter on response characteristics and to validate the results. A confirmation test was executed.

ACKNOWLEDGMENTS

The authors would like to thank Dr. Subhash Singh (assistant faculty member, NIT Jamshedpur, Jharkhand, India) for adding to this chapter and for giving technical support.

REFERENCES

Dhar S., Purohit R., Saini N., Sharma A., and Kumar G.H., 2007. Mathematical modeling of electric discharge machining of cast Al–4Cu–6Si alloy–10 wt.% SiCP composites, *Journal of Materials Processing Technology* 194(1–3), 24–29.

Eubank P.T., Patel M.R., Barrufet M.A., and Bozkurt B., 1993. Theoretical models of the electrical discharge machining process. III. The variable mass, cylindrical plasma model, *Journal of Applied Physics* 73(11), 7900–7909.

Fattouh M., Elkhabeery M., and Fayed A.H., 1990. Modelling of some response parameters in EDM, AME Fourth Conference Military Technical College, Cairo, Egypt.

Hung N.P., Yang L.J., and Leong K.W., 1994. Electrical discharge machining of cast metal matrix composites, *Journal of Materials Processing Technology* 44(3–4), 229–236.

Khan M.Y., and Rao P.S., 2019. Electrical discharge machining: vital to manufacturing industries, *International Journal of Innovative Technology and Exploring Engineering* 8(11), 1696–1701.

Khan M.Y., Rao P.S., and Pabla B.S., 2020a. Investigations on the feasibility of jatropha curcas oil based biodiesel for sustainable dielectric fluid in EDM process, *Materials Today: Proceedings* 26, 335–340.

Khan M.Y., Rao P.S., and Pabla B.S., 2020b. Powder mixed electrical discharge machining (PM-EDM): a methodological review, *Materials Today: Proceedings*. (Available online).

Khan M.Y., Rao P.S., and Pabla, B. S., 2020c. Review of electrical discharge machining process with nanopowder and CNT mixed dielectric fluid, Proceedings of 4th International Online Multidisciplinary Research Conference (IOMRC-2020), Hyderabad, India.

König W., Dauw D.F., Levy G., and Panten U., 1988. EDM-future steps towards the machining of ceramics, *CIRP Annals* 37(2), 623–631.

Lau W.S., Yue T.M., Lee T.C., and Lee W.B., 1995. Un-conventional machining of composite materials, *Journal of Materials Processing Technology* 48(1–4), 199–205.

Müller F., and Monaghan J., 2000. Non-conventional machining of particle reinforced metal matrix composite, *International Journal of Machine Tools and Manufacture* 40(9), 1351–1366.

Müller F., and Monaghan J., 2001. Non-conventional machining of particle reinforced metal matrix composites, *Journal of Materials Processing Technology* 118(1–3), 278–285.

Natsu W., Shimoyamada M., and Kunieda M., 2006. Study on expansion process of EDM arc plasma, *JSME* 49(2), 600–605.

Panda D.K., 2008. Study of thermal stresses induced surface damage under growing plasma channel in electro-discharge machining, *Journal of Materials Processing Technology* 202(1–3), 86–95.

Rizwee M., Minz S.S., Orooj M., Hassnain M.Z., and Khan M.J., 2019. Electric discharge machining method for various metal matrix composite materials, *International Journal of Innovative Technology and Exploring Engineering* 8(9), 1796–1807.

Rizwee M., and Rao P.S., 2018. A review on electro discharge machining of metal matrix composite, *International Journal of Technical Innovation in Modern Engineering & Science* 4(12), 410–414.

Rizwee M., and Rao P.S., 2021. Analysis & optimization of parameters during EDM of aluminium metal matrix composite, *Journal of University of Shanghai for Science and Technology* 23(3), 218–223.

Rizwee M., Rao P.S., and Khan M.Y., 2021. Recent advancement in electric discharge machining of metal matrix composite materials, *Materials Today: Proceedings* 37(2), 2829–2836.

Rao P.S., Dwivedi D.K., and Jain P.K., 2014. Study and prediction of surface finish of external cylindrical surfaces of titanium alloy by electro chemical honing (ECH) process, 23rd International Conference on PFAM-2014 at IITR, Roorkee, India.

Rao P.S., Dwivedi D.K., and Jain P.K., 2015. Study and effect of process parameters on external cylindrical surfaces of titanium alloy by electro chemical honing (ECH) process, 26th DAAAM International Symposium on Intelligent Manufacturing and Automation 2015, Zadar, Croatia.

Rao P.S., Dwivedi D.K., and Jain, P.K., 2016. Prediction of optimal process parameters on to electro chemical honing (ECH) of external cylindrical surfaces of titanium alloy using doe technique, 6th International and 27th All India Manufacturing Technology, Design and Research Conference (AIMTDR 2016) at COEP, Pune, India.

Rao P.S., Dwivedi D.K., and Jain P.K., 2017. Optimization of key process parameters on electro chemical honing (ECH) of external cylindrical surfaces of titanium alloy Ti-6Al-4V, *Materials Today: Proceedings* 4(2), 2279–2289.

Senthilkumar V., and Omprakash B.U., 2011. Effect of titanium carbide particle addition in the aluminium composite on EDM process parameters, *Journal of Manufacturing Processes* 13(1), 60–66.

Singh P.N., Raghukandan K., Rathinasabapathi M., and Pai B.C., 2004. Electric discharge machining of Al–10% SiCP as-cast metal matrix composites, *Journal of Materials Processing Technology* 155, 1653–1657.

8 Rheological Performance of Coconut Oil-based Mono- and Bi-dispersed Magnetorheological Finishing Fluid (MRFF)

Vikas Kumar, Rajesh Kumar, and Harmesh Kumar

CONTENTS

8.1 INTRODUCTION

The finishing performance of the magnetorheological abrasive finishing (MRAF) process mainly depends on the machine parameters and finishing medium composition (Kumar et al., 2020). The finishing medium employed in the MRAF process is generally termed as "magnetorheological finishing fluid" (MRFF). Along with machine parameters, MRFF generates a sufficient amount of finishing forces to remove the surface asperities of the work sample. Many researchers have reported on rheological studies and issues of sedimentation problems in the general composition of MR fluids. However, the rheology of such fluid becomes complex when nonmagnetic abrasives are added to it (Jha and Jain, 2009; Sidpara et al., 2009). Such unique findings are subject to further investigation. When some magnetic field is applied to an MRFF sample, it transforms into a stiffened structure due to the alignment of micron-sized magnetizable particles along the magnetic field lines. In doing so, the viscosity and yield strength of an MRFF sample increases to a great extent. However, it can be further improved using nano-sized magnetizable particles along with micron-sized particles. The stiffened structure of MRFF possesses a certain number of microcavities in between the aligned chain

DOI: 10.1201/9781003203681-10

of micron-sized particles. The addition of nano-sized particles along with micron-sized particles may fill the microcavities of the aligned microparticles, thus, forming a highly compact and rigid structure under the effect of applied magnetic field (Niranjan and Jha, 2014; Leong et al., 2016). Moreover, the nanoparticles have a high surface area to volume ratio due to which they possess excellent dispersibility in the suspending medium.

The environment-friendly nature of MRFFs is equally important with their simultaneous use in finishing applications. It is mainly dependent on the selection of MRFF constituents (Kumar et al., 2019). From the literature, it is revealed that silicon oil, mineral oil, hydrocarbon oil, deionized water, etc., are the commonly used carrier liquids used for MRFF synthesis. Among the aforementioned carriers, the aqueous medium is found to be quite promising because of its environment-friendly nature. However, the use of deionized water possesses certain problems. Keeping in view the previously mentioned problems and the issue of environment-friendliness, an eco-friendly carrier liquid and additives can be used to synthesize MRFF samples. Therefore, the present work was carried out to synthesize environment-friendly MRFF samples using a coconut oil-based carrier liquid with the simultaneous aim to improve their rheological performance using a mixture of micro-sized and nano-sized magnetizable particles.

8.2   MATERIAL SELECTION

This section deals with various constituents of MRFF, which includes micron-sized and nano-sized magnetizable particles, abrasives, and carrier liquid and additives. Coconut oil (environment-friendly oil), which was procured from the local market, was used as a carrier liquid medium. The guar gum was used as a thixotropic additive, which is a rigid, nonionic, polydisperse neutral carbohydrate polymer. The micron-sized Fe particles with a spherical shape were procured from the Sigma-Aldrich Corporation. The magnetite Fe_3O_4 with a mean particle size of 50–100 nm were procured from the Sigma-Aldrich Corporation. The silicon carbide (SiC) abrasive powder was procured from Speedfam Co. Ltd.

The micron-sized Fe particles, nano-sized Fe_3O_4 particles, and SiC abrasives were characterized using the of X-ray diffraction (XRD) technique. The obtained XRD pattern of micron-size Fe particles, nano-sized Fe_3O_4 particles, and SiC abrasives are shown in Figures 8.1–8.3, respectively.

8.3   SYNTHETIZATION OF MRFF SAMPLES

The micro-nano magnetizable particle-based bidispersed MRFF sample was synthesized by adding 5 vol% concentration of nano-sized Fe_3O_4 particles within a fixed 35 vol% concentration of micron-sized Fe particles. The remaining constituents include 10% volume fraction of SiC abrasives and 55% volume fraction of base fluid medium. The monodispersed MRFF sample was also synthesized to compare the rheological performance of both the MRFF samples. The composition of the monodispersed MRFF sample includes 35% volume fraction of micron-sized Fe particles, 10% volume fraction of SiC abrasives, and 55% volume fraction

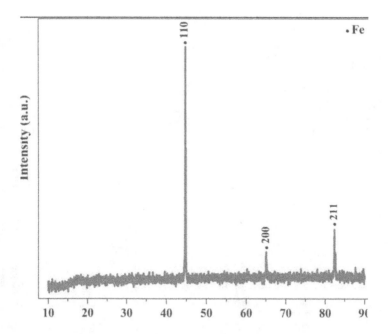

FIGURE 8.1   XRD pattern of spherically shaped Fe particles.

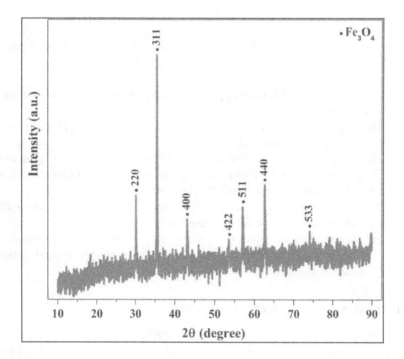

FIGURE 8.2   XRD pattern of nano-sized Fe_3O_4 particles.

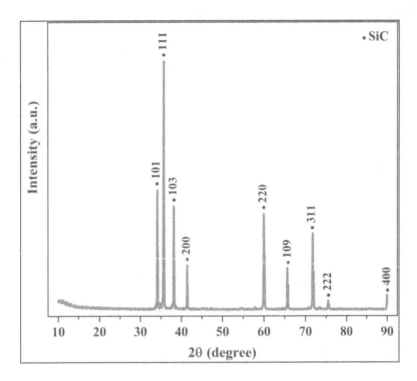

FIGURE 8.3   XRD pattern of SiC abrasives.

of base fluid medium. The following steps were followed to synthesize both the MRFF samples:

- First, base fluid medium was prepared using organic phase solution and aqueous phase solution.
- The organic phase solution was prepared by the intermixing of the emulsifying agent with coconut oil.
- The water phase solution was prepared by first mixing the guar gum and glycerin together and then by trickling the deionization (DI) water while mixing was done using a stirrer for about 20 minutes.
- Both of the phases were further intermixed with the help of a stirrer for another 20 minutes to obtain the base fluid medium.
- Finally, the MRFF samples were prepared by mixing all the constituents (i.e., magnetizable particles, SiC abrasives, and base fluid medium) together for about 20 minutes.

8.4   MAGNETORHEOLOGICAL TESTING

To compare the rheological behavior of both MRFF samples, the rheological testing was done using a rheometer MCR-102 model equipped with an MRD 180 unit (Anton Paar, Graz, Austria). The MRD 180 unit delivers the homogeneous magnetic field at right angles to the shear flow direction of the MRFF sample. To perform rheological testing, the parallel plate geometry was used with a 0.1-cm working slit arrangement and the temperature condition was maintained to 25°C. Finally, the rheological testing was performed at low (i.e., 0.2 T) and high (i.e., 0.8 T) magnetic field over the shear rate range increased up to 1000 s^{-1}.

8.5 RESULTS AND DISCUSSION

The comparative rheological analysis of monodispersed and bidispersed MRFF samples was done on the basis of the results obtained from magnetorheological testing, which was performed at low (i.e., 0.2 T) and high (i.e., 0.8 T) magnetic field. The testing results were obtained as shear stress versus shear rate and viscosity versus shear rate.

Figure 8.4(a) and (b) illustrates the variation in shear stress with respect to the shear rate of both MRFF samples tested at 0.2 and 0.8 T field strength, respectively. The corresponding change in viscosity with respect to shear rate of both MRFF samples at 0.2 and 0.8 T is shown in Figure 8.5(a) and (b), respectively. Figure 8.4(a) illustrates that the flow curve of the bidispersed MRFF sample leveled up over the entire range of shear rate vis-à-vis monodispersed MRFF. This is because the bidispersed MRFF exhibits higher viscosity over the entire range of shear rate as construed from Figure 8.5(a), and is attributed to filling the microscopic voids in between micron-sized particles by the Fe_3O_4 nanoparticles. This results in a more compact and rigid assembly of particles under applied magnetic field. At 0.2 T, the magnetic moment of nanometer Fe_3O_4 particles contributes to an increase in the overall magnetic strength of the bidispersed MRFF stiffened under the applied field. In contrast, the flow curve of the monodispersed MRFF sample leveled up for the entire shear rate region compared with the bidispersed MRFF tested under 0.8 T (Figure 8.4(b)). At high magnetic field (i.e., 0.8 T), nanometer Fe_3O_4 particles do not contribute to an increase in the overall magnetic strength of the bidispersed MRFF. This is because the magnetic saturation level of nanometer Fe_3O_4 particles is very low compared with the micron-sized iron particles and that is why they do not contribute much at a high magnetic field of 0.8 T.

8.6 CONCLUSIONS

The present work deals with the synthesis of the MRFF sample using coconut oil-based carrier liquid by considering the importance of the environment-friendly nature of MRFF. The magnetorheological results show that the coconut oil can be used to synthesize the MRFF samples. At low magnetic field, the bidispersed MRFF sample shows better rheological performance vis-à-vis monodispersed MRFF, as the Fe_3O_4 nanoparticles fill the microscopic voids and contribute to increase the overall magnetic strength of the bidispersed MRFF. In contrast, the bidispersed MRFF does not perform well at high magnetic field due to the low saturation magnetization of the Fe_3O_4 nanoparticles.

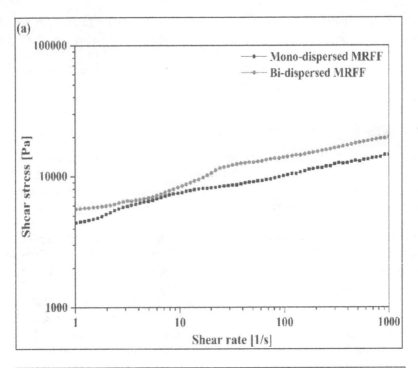

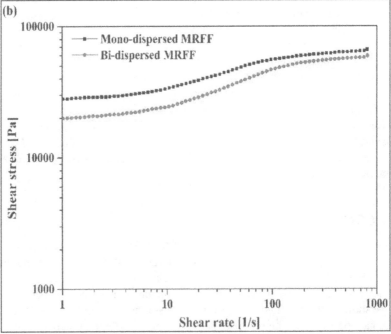

FIGURE 8.4   Magnetorheological testing results of shear stress versus shear rate at (a) 0.2 T and (b) 0.8 T.

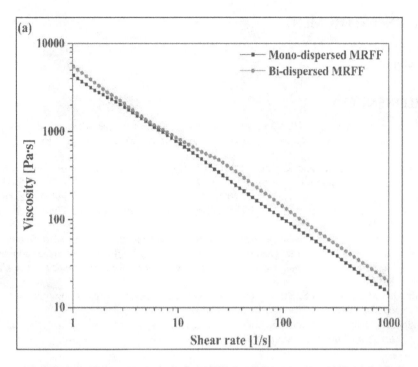

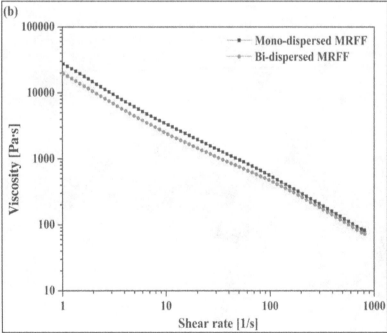

FIGURE 8.5 Magnetorheological testing results of viscosity versus shear rate at (a) 0.2 T and (b) 0.8 T.

ACKNOWLEDGMENTS

The authors thank I.I.T., Ropar, and U.I.E.T., Panjab University, Chandigarh, for lab facilities.

REFERENCES

Jha S., and Jain V.K., 2009. Rheological characterization of magnetorheological polishing fluid for MRAFF, *International Journal of Advanced Manufacturing Technology* 42, 656–668.

Kumar V., Kumar R., and Kumar, H., 2019. Rheological characterization of vegetable-oil-based magnetorheological finishing fluid, *Materials Today: Proceedings* 18, 3526–3531.

Kumar V., Kumar R., and Kumar, H., 2020. Rheological characterization and performance evaluation of magnetorheological finishing fluid, *Journal of Applied Fluid Mechanics* 13, 185–197.

Leong S.A.N., Samin P.M., Idris A., Mazlan S.A., and Rahman, A.H.A., 2016. Synthesis, characterization and magnetorheological properties of carbonyl iron suspension with superparamagnetic nanoparticles as an additive, *Smart Materials and Structures* 25, 1–12.

Niranjan M.S., and Jha S., 2014. Flow behaviour of bidisperse MR polishing fluid and ball end MR finishing *Procedia Materials Science* 6, pp. 798–804.

Sidpara A., Das M., and Jain V.K., 2009. Rheological characterization of magnetorheological finishing fluid, *Materials and Manufacturing Processes* 24, 1467–1478.

9 Enhancing Different Weld Attributes of Ultrasonically Welded Al/Ni Joints for Battery Electric Vehicles

Soumyajit Das, Mantra Prasad Satpathy,
Bharat Chandra Routara,
and Susanta Kumar Sahoo

CONTENTS

9.1 INTRODUCTION

Current upsurge in automotive technologies and demand for electric vehicles compels the manufacturer to meet the environmental challenges and improve fuel efficiency. Moreover, battery electric vehicles (BEVs) are increasingly being used to reduce human death-causing emissions (Seo et al., 2015; Anenberg et al., 2017; Rockström et al., 2017). Meanwhile, the battery packs present inside these vehicles are composed of a large number of battery cells that are electrically connected to transmit the right amount of power. Aluminum (Al) and nickel (Ni) alloys are the two most preferred electrically conductive materials in the lithium-ion batteries of BEVs (McNutt, 2013). However, there are some difficulties in joining these dissimilar alloys by fusion welding processes because of the differences in their physical, metallurgical, and chemical properties. Furthermore, the detrimental intermetallic compound (IMC) formation during the conventional fusion welding process extremely weakens the joint strength (Matsuoka and Imai, 2009). To overcome this problem, several

DOI: 10.1201/9781003203681-11

researchers are now focused on solid-state joining processes like diffusion welding (Bang et al., 2013; Li et al., 2019), explosive welding (Mahendran et al., 2010), friction stir welding (Yan et al., 2010; Çam, 2011), and ultrasonic spot welding (USW; Ipekouglu et al., 2013; Peng et al., 2019), which produce superior joints with few IMC formations. Among these techniques, USW is an energy-efficient solid-state method implied for joining non-ferrous metals without melting or adding any filler materials. It allows joints to have high static properties and low through-thickness porosity compared with other solid-state welding processes. From a production viewpoint, the USW system has a high potential for industrialization due to a shorter weld cycle (<1 s) and no shrinkage nature for small joining areas (Das et al., 2019). Throughout the USW process, a specimen is clamped between the sonotrode and anvil with a certain amount of weld pressure (WP). The sonotrode part vibrates parallelly to the weld spot by using high-frequency ultrasonic vibrations (15–40 kHz) to produce a bond formation at the faying surface of the specimen. A schematic diagram of the standard USW process is shown in Figure 9.1. This process is most appropriate for joining thin components like wires, medical instruments, and packaging materials (Fujii et al., 2018; Ao et al., 2019).

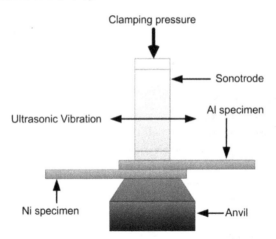

FIGURE 9.1 Schematic diagram of USW.

Numerous researchers studied the joint mechanism, weld strength, consistency of ultrasonic welds, and optimum parametric conditions to improve the bonding formation and decrease IMC thickness during the USW process. Hetrick et al. (2009) tested with AA6111 and AA5754 sheets to characterize the outcomes of USW process parameters on the weld interfacial area. The result revealed that no noticeable variation was seen on the tensile shear and T-peel test. Moreover, the different weld conditions disclosed the relationship between weld quality and mechanical performance of the ultrasonic spot welded joints. Daniels (1965) studied different types of weld factors like pressure, weld duration, and power to discuss material bonding phenomena between Al-Cu joints. The result showed that the physical properties and the thickness of the welded specimens were drastically changed due to several welding conditions. The surface conditions of the welded materials also played an important role throughout the USW process. Kim et al. (2011) performed the ultrasonic welding process by joining the Ni-plated Cu and Cu sheet to identify the various failure natures in T-peel tests. Peng et al. (2017) investigated the changes of mechanical and microstructural properties of ultrasonic spot welded Al 6022 alloys at various weld energies. The results reported that the equiaxed grain structure was noticed at maximum weld energy (WE). Ni and Ye (2016) analyzed the joining mechanism between the Al and Ni sheets with an interlayer and without an interlayer for evaluating the weld quality. As a result, several weld attributes were observed to improve the temperature of the weld region by using an interlayer.

The objective of this chapter is to systematically investigate the effects of different weld parameters on 0.3-mm Al (AA1100) and Ni sheet joints by using the USW technique. The joint strength like lap shear tensile properties and the Vickers microhardness at the weld cross section are discussed in this chapter. Furthermore, to improve the quality of the weld joints, the various microstructural evolution features at the weld interface are revealed.

9.2   EXPERIMENTAL DETAILS

Commercial Al (AA1100) sheets and pure Ni sheets with a thickness of 0.3 mm are used in this study. The specimen's dimensions are 75-mm length and 20-mm width. These were cut from the large metal sheets along the rolling direction for preparation of the weld samples. Mechanical properties of these two base materials are listed in Table 9.1. The joining surfaces of both base metals were grounded with 800- and 1200-grit sandpapers followed by cleaning with acetone and dried before the welding process. Throughout the welding process, the Ni specimen was kept at the anvil side, whereas Al was placed on top of the Ni surface. For all the configurations, a 20-mm overlap of Al and Ni specimens was ensured. A schematic diagram of the lap configured welding specimen is shown in Figure 9.2(a).

TABLE 9.1
Mechanical Properties of Aluminum (AA1100) and Nickel Alloy

Materials	Ultimate Tensile Strength (MPa)	Yield Strength (MPa)	Thermal Conductivity (Wm^{-1}°C^{-1})	Specific Heat (Jg^{-1}°C^{-1})
AA1100	89.6	20	222	0.904
Nickel	317	59	60.70	0.460

The Telsonic® M4000 lateral drive USW machine was utilized to perform the experiments with the maximum output power of 3000 W and the frequency of 20 kHz. A complete setup of the USW system is demonstrated in Figure 9.2(b). The dimensions of the sonotrode tip are 9 × 11 mm with a height of 0.3 mm, as shown in Figure 9.2(c). The direction of the vibration is perpendicular to the rolling direction of the sheets. The experimental work was carried out with the changes of different weld conditions such as WP and WE at the vibrational amplitude (A) of 47 μm. The weld

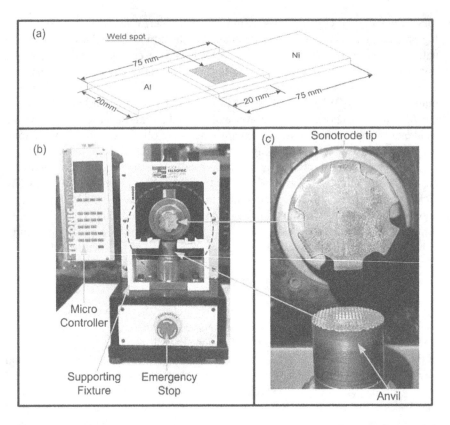

FIGURE 9.2 (a) Schematic diagram of welding specimen, (b) USW machine, and (c) Sonotrode tip and anvil surface.

conditions with their levels are shown in Table 9.2. The tensile lap shear test was performed at strain rate of 2 mm/min by applying a 20-kN load in an Instron® 3300 machine. A digital Vickers hardness tester was used to determine the material softening phenomena at the weld interface. An indentation load of 450 gf with a dwell time of 15 s was employed for each specimen. The fracture morphology and various microstructural conditions of ultrasonic spot welded joints were investigated by

TABLE 9.2
Several Levels of Weld Parameters

Factors	Units	Level 1	Level 2	Level 3	Level 4	Level 5	Level 6
Amplitude	μm	47	...	...	...	...	...
Weld pressure	MPa	0.22	0.24	0.26	...	...	...
Weld energy	J	210	420	525	630	735	840

field emission scanning electron microscopy (FESEM) and energy dispersive X-ray (EDX) analyses.

9.3   RESULTS AND DISCUSSION

9.3.1   TENSILE LAP SHEAR STRENGTH

The tensile lap shear strength result indicates the relationship between bonding strength and WE. The effect of WE and WP at a constant amplitude of 47 μm on the tensile strength is demonstrated in Figure 9.3. It shows that the joint strength gradually increased with the increase of WE up to 630 J at 0.24 MPa of WP and achieved the highest value of 452 N. Furthermore, as the WE increased longer than 630 J, severe plastic deformation was noticed at the weld spot because of the synergetic effect of vibration and compression applied from the sonotrode. Under this situation, the Al-Ni joint severely ruptured and adhered to the sonotrode. Thus, it reveals that the increase of WP and WE can cause more relative motion between the weld specimens, eventually leading to a massive amount of interface temperature generation. As a result, more plastic deformation and brittle IMCs are formed at the weld area, which reduces the bond strength of the Al-Ni joints.

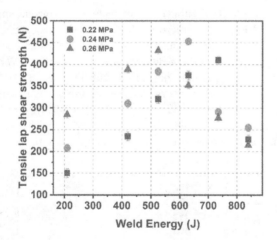

FIGURE 9.3   Tensile lap shear strength of Al-Ni welded specimens for various WP and WE.

Figure 9.4 portrays the relationship between tensile lap shear strength and weld interface temperature with the static WP of 0.24 MPa at fixed vibration amplitude. By comparing the weld interface temperature with the tensile shear strength, it was noticed that tensile shear strength increased with the rise in weld interface temperature. As the interfacial temperature increased, the plastic deformation of Al became easier and more an enhanced bond strength was achieved. In this experiment, the tensile lap shear strength increased up to 630 J of WE.

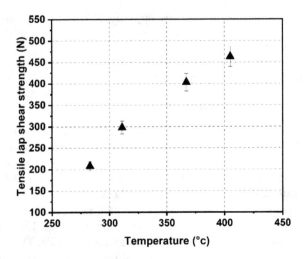

FIGURE 9.4 Tensile lap shear strength of Al-Ni welded specimens for different weld temperatures.

9.3.2 MICRO-HARDNESS ANALYSIS

To investigate the material softening phenomena for the ultrasonic spot welded Al-Ni joints at different WE, Vickers micro-hardness analysis was carried out. Figure 9.5 shows the variation of hardness values on the cross-sectional area of the weld specimens at an equidistance from the weld interface at 0.24 MPa of WP. It was noticed that the hardness value gradually decreased with the increase of WE due to the changes in weld interface temperature. As a result, grain-sized changes and recrystallization occur in the weld interface region. At the WE of 840 J, the minimum hardness value was observed at the center of the weld spot. Therefore, the hardness value at the center of the cross section is lower compared with the original sheet.

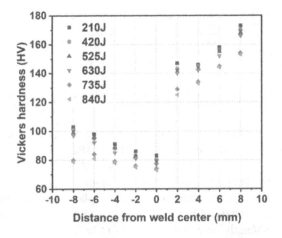

FIGURE 9.5 Hardness deviation cross-sectionally with different weld energy.

9.3.3 Microstructural Analysis

To evaluate the bonded Al-Ni specimen microstructural characteristics, FESEM images were obtained and compared along with EDS line scan analysis. The FESEM results along the bond line of Al-Ni joints at different WE and WP are demonstrated in Figure 9.6(a–c). The unbonded area at the weld interface indicates that the Al-Ni specimens are not properly welded when the WE is 210 J and the WP is 0.22 MPa (Figure 9.6(a)). As the WE increased to 630 J, the gap between the Al-Ni sheets disappeared at a WP of 0.24MPa (Figure 9.6(b)). Moreover, an internal bonding with a minor round-type void is formed without any gaps at the Ni side. The wavy-like interface pattern reveals that the weld region is affected by the WE of 840 J at a WP of 0.26 MPa in Figure 9.6(c). In fact, many microcracks were noticed at the Al surface because the penetration of the sonotrode increased with the increase in WE across the weld spot. It is noted that several IMC bond formations at the weld interface are beneficial for joint strength, but excessive plastic deformation may hamper the weld quality of the joint as well as weld durability.

Furthermore, to differentiate the weld attributes more diligently, an EDS line scan analysis was executed on the weld interface. Figure 9.6(d–f) shows the EDS line scan images to confirm the interdiffusion thickness and possible IMC phases formed during the USW process. Figure 9.6(d) indicates that the diffusion bond area is almost negligible due to insufficient plastic deformation and the small amount of material that flow occurs at the weld region. As the weld energy and pressure increased, these unbonded gaps were completely removed and a thick interdiffusion zone was clearly noticeable around the weld area (Figure 9.6(e)). In Figure 9.6(f), it is exposed that a lamellar crack is detected at the Al surface due to extreme atomic diffusion. From the EDS line scan, it can be concluded that a constant microscale transition phase was formed at the weld region, containing Al and Ni parent elements. It was

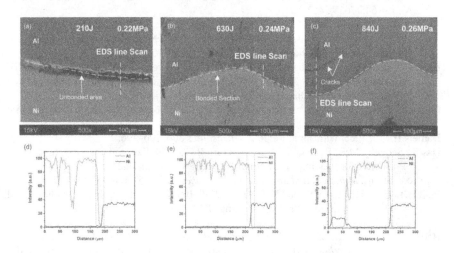

FIGURE 9.6 (a–c) FESEM images showing different weld qualities and (d–f) EDS line scan results for different welded specimens.

confirmed that weld energy and pressure play an important role in the joining of Al-Ni sheets during this USW process.

9.4 CONCLUSIONS

The current study successfully analyzed the different mechanical attributions and microstructural changes of Al (AA1100)-Ni dissimilar sheet metals by using the USW process. The following conclusions could be drawn:

- The maximum tensile lap shear strength value was obtained as 452 N at 0.24 MPa of WP and 630 J of WE. Subsequently, the weld strength was decreased as the weld energy and pressure further increased.
- The micro-hardness analysis showed that the hardness value was minimum at the center of the weld cross-sectional area, which was lower than the base metal. This generally happened because of an excessive amount of cold working occurred in the weld region.
- The microstructural analysis uncovered several weld strength qualities that depended on the WE and WP values at a fixed vibration amplitude of 47 μm. A swirl-like interface pattern was noticed when the weld spot was affected by the WE. The bonding mechanism confirmed that too much plastic deformation might weaken the weld strength. Additionally, the EDS line scan disclosed the interdiffusion nature and the formation of cracks at the weld interface.

REFERENCES

Anenberg S.C., Miller J., Minjares R., Du L., Henze D.K., Lacey F., Malley C.S. et al., 2017. Impacts and mitigation of excess diesel-related NO_x emissions in 11 major vehicle markets, *Nature* 545(7655), 467–471.

Ao S., Li C., Zhang W., Wu M., Dai Y., Chen Y., and Luo Z., 2019. Microstructure evolution and mechanical properties of Al/Cu ultrasonic spot welded joints during thermal processing, *Journal of Manufacturing Processes* 41, 307–314.

Bang H.S., Bang H.S., Song H.J., and Joo S.M., 2013. Joint properties of dissimilar Al6061-T6 aluminum alloy/Ti–6% Al–4% V titanium alloy by gas tungsten arc welding assisted hybrid friction stir welding, *Materials & Design* 51, 544–551.

Çam G., 2011. Friction stir welded structural materials: beyond Al-alloys, *International Materials Reviews* 56(1), 1–48.

Daniels H.P.C., 1965. Ultrasonic welding, *Ultrasonics* 3(4), 190–196.

Das S., Satpathy M.P., Pattanaik A., and Routara B.C., 2019. Experimental investigation on ultrasonic spot welding of aluminum-cupronickel sheets under different parametric conditions, *Materials and Manufacturing Processes* 34(15), 1689–1700.

Fujii H.T., Endo H., Sato Y.S., and Kokawa H., 2018. Interfacial microstructure evolution and weld formation during ultrasonic welding of Al alloy to Cu, *Materials Characterization* 139, 233–240.

Hetrick E.T., Baer J.R., Zhu W., Reatherford L.V., Grima A.J., Scholl D.J., Wilkosz D.E., Fatima S., and Ward S.M., 2009. Ultrasonic metal welding process robustness in aluminum automotive body construction applications, *Welding Journal* 88(2), 149–158.

Ipekouglu G., Erim S., Gören-Kiral B., and Çam G. 2013. Investigation into the effect of temper condition on friction stir weldability of AA6061 Al-Alloy plates, *Kovove Materialy* 51(3), 155–163.

Kim T.H., Yum J., Hu S.J., Spicer J.P., and Abell J.A., 2011. Process robustness of single lap ultrasonic welding of thin, dissimilar materials, *CIRP Annals* 60(1), 17–20.

Li F., Liu T., Zhang J.Y., Shuang S., Wang Q., Wang A.D., Wang J.G., and Yang Y., 2019. Amorphous–nanocrystalline alloys: fabrication, properties, and applications, *Materialstoday Advances*. https://doi.org/10.1016/j.mtadv.2019.100027

Mahendran G., Babu S., and Balasubramanian V., 2010. Analyzing the effect of diffusion bonding process parameters on bond characteristics of Mg-Al dissimilar joints, *Journal of Materials Engineering and Performance* 19(5): 657–665.

Matsuoka S., and Imai H., 2009. Direct welding of different metals used ultrasonic vibration, *Journal of Materials Processing Technology* 209(2), 954–960.

McNutt M., 2013. Climate change impacts, *Science* 341(6145), 435.

Ni Z.L., and Ye F.X., 2016. Weldability and mechanical properties of ultrasonic welded aluminum to nickel joints, *Materials Letters* 185, 204–207.

Peng H., Chen D.L., Bai X.F., She X.W., Li D.Y., and Jiang X.Q., 2019. Ultrasonic spot welding of magnesium-to-aluminum alloys with a copper interlayer: microstructural evolution and tensile properties, *Journal of Manufacturing Processes* 37, 91–100.

Peng H., Chen D., and Jiang X., 2017. Microstructure and mechanical properties of an ultrasonic spot welded aluminum alloy: the effect of welding energy, *Materials* 10(5), 449.

Rockström J., Gaffney O., Rogelj J., Meinshausen M., Nakicenovic N., and Schellnhuberm H.J., 2017. A roadmap for rapid decarbonization, *Science* 355(6331), 1269–1271.

Seo J.S., Jang H.S., and Park D.S., 2015. "Ultrasonic welding of Ni and Cu sheets, *Materials and Manufacturing Processes* 30(9), 1069–1073. doi: https://doi.org/10.1080/10426914.2014.880466

Yan Y.B., Zhang Z.W., Shen W., Wang J.H., Zhang L.K., and Chin B.A., 2010. Microstructure and properties of magnesium AZ31B–aluminum 7075 explosively welded composite plate, *Materials Science and Engineering: A* 527(9), 2241–2245.

10 Optimization by Grey Relational Analysis of AWJM Parameters on Machining Titanium Alloy

Chandrakant Chaturvedi, P. Sudhakar Rao, and Mohd. Yunus Khan

CONTENTS

10.1 INTRODUCTION

Throughout the world, nonconventional machining processes are gaining significance due to several merits. Effectual usage of the capability of various nonconventional machining methods requires careful selection of the appropriate technique for a given application (Rao et al., 2014a,b, 2015a,b, 2016a,b, 2017; Khan and Rao, 2019a,b; Khan et al. 2020). Water jet technology has evolved from plain water jet cutting and has an application in a variety of industries. Mechanically linked industries such as automobile and aerospace find it suitable in many applications, apart from the construction industry, which also applies this technique for specially designed equipment and sections. This technology is also being applied in the chemical process industry and industrial maintenance work (Momber and Kovacevic, 2012). Initially, pure water jet technique was applied, but it was not suitable for hard material, and pure water jet was employed with an abrasive technique and was able to process many types of material processing that were not possible with other conventional and nonconventional techniques. A combined abrasive and water jet system is known as abrasive water jet machining (AWJM; Senthilkumar et al., 2020). Many researchers have performed experimental studies on AWJM using different materials considering a distinct set of input and response parameters. Senthilkumar et al. (2020) investigated the effect of various input variables, including pressure, stand-off distance (SOD) and feed rate on the kerf angle, erosion rate, and surface roughness (SR) of a machined surface created after AWJM of a hybrid metal matrix

DOI: 10.1201/9781003203681-12

composite (HMMC). Aluminum-based HMMCs were machined by AWJ as per the L_{27} orthogonal array. Obtained results were considered and examined with ANOVA and the Grey relational analysis (GRA) technique. GRA is applied to obtain multi-objective optimum results. Researchers found a combination of optimum input variables (350 bar pressure, 10 mm/min speed, and 1 mm SOD) for improved responses. Joel and Jeyapoovan (2020) obtained hardness, material removal rate (MRR), and SR at various sets of experiments created in Taguchi design considering abrasive flow rate (AFR), SOD, and nozzle speed as varying parameters. Obtained results were considered for multioptimization using the GRA technique to overcome the limitation of Taguchi analysis. Obtained AFR and nozzle speed significantly improved response variables. All the experimental work is performed on AA7075 aluminum alloy. Naresh Babu and Muthukrishnan (2015) considered multiple input variables that were taken from various categories of machining parameters, such as mesh size and AFR taken from the abrasive section, pressure from the machining section, nozzle diameter is taken from the designed parameters and SOD, and feed rate taken from the machining section. Nair and Kumanan (2018) considered multiple response factors in the AWJ drilling process. Machining work was performed on Inconel 617 and the drilling rate, SR, circularity, and taper angle was measured.

Garnet and aluminum oxide are most commonly used abrasive particles in AWJM. Apart from said materials, sand and glass globules are also utilized for their scratchiness. The function of the abrasive particles is to enhance the cutting ability of the water jet (Hong et al., 2001; Seo et al., 2003). Titanium and its alloys are known to be difficult machining materials and these alloys consist of low machinability due to poor thermal conductivity and chemical reactivity with the materials from which tools are prepared. Ti-6Al-4V alloys are one of the phases consisting of an alpha and beta mixture (Seo et al., 2003). Considering specific properties of the alloy, it is applied in aeronautical engine elements, medical equipment, structures, and bioimplants. Among numerous alloying types, Ti-6Al-4V is the maximum used structural alloy in the aeronautical industry, providing 45% of the industry's whole titanium alloy consumption (Ducato et al., 2013). One merit of AWJM is that its performance does not depend on the thermal and chemical properties of the material. Hence, in this study, the machining experimentation of titanium is performed on AWJM. The parameters are selected from previous research work and their future scope.

10.2 MATERIALS AND METHODS

Ti-6Al-4V titanium alloy has been machined by AWJ and reported by many researchers; it is also extensively used in various industrial applications. In this work, a $10 \times 145 \times 150$-mm^3 plate of titanium alloy is used for experimental work. The metal is purchased from Dali Electronics, Mumbai, India. The purchased material was tested using an energy dispersive spectroscope (EDS).

Table 10.1 shows the chemical composition of material tested. Elements present in the material are Al (6.01%), V (4%), Fe (0.11%), C (0.03%), N (0.018%, H (0.008%), and O (0.1%).

The AWJM (model OMAX 5555) with a table size of 5×5 and pressure capacity of 55,000 psi, which is available at TECHAIDS, Mohali (Punjab), India, was used

TABLE 10.1

Chemical Composition of Material

Element	Ti	Al	V	Fe
Percentage (by weight)	90.79	6.96	2.12	0.13

to machine selected material for experimentation. The experiment was performed as per design of experiment (DOE) obtained using Taguchi $L_{25}(5^4)$. Value input variables used to design the experiment include pressure (25, 30, 35, 40, and 45 kpsi) AFR (300, 400, 500, 600, and 700 g/min) SOD (2, 4, 6, 8, 10), and quality (1–5).

Optimization is the technique of acquiring the best outcomes in given conditions. There are several techniques of optimization used in the machining process (Rao et al., 2014a,b; 2015a,b). Researchers have used this technique to limit the trials for their experiments by creating the orthogonal array (OA), which is further used for statistical analysis and prediction of results (Ross, 1996; Roy, 2001; Kumar et al., 2021)

In this work, the GRA is considered to resolve the problem associated with the previous technique. The word Grey is associated with incomplete, suspicious information. Analysis of such information is not possible with traditional statistical methods; hence, Grey theory provides a significant solution to examine such doubtful information or collected data (Lin and Ho, 2003; Kumar et al., 2020). GRA is applied to optimize machining time (MT), SR, and hardness (HRC) considering pressure, SOD, AFR, and quality as machining control variables.

10.3 RESULTS AND DISCUSSION

Response variables were measured by suitable instruments and a corresponding signal-to-noise (S/N) ratio was calculated using Minitab 18 software. Table 10.2 presents the experimental results corresponding to S/N ratio, Grey relation coefficient (GRC), and Grey relation grade (GRG), along with the ranks of the concerned set of variables.

GRG results are statically analyzed to prepare the response table presented in Table 10.3. The aim of the analysis is to obtain the significance of all the selected input parameters on the responses. Table 10.3 also provides the optimum combination of variables such as pressure at level 1, AFR at level 3, SOD at level 4, and quality 5. Pressure is the most impactful parameter, whereas SOD has the least impact on the response. Further, ANOVA was carried out to check the validity of the obtained results, which are found to be 97.67% correct and suit the criteria of the experimentation work.

Percentage contribution of each variable is obtained by performing where pressure has the maximum contribution of 56.44% and SOD has the least at 4.08%. P value for SOD is more than 0.05, which shows the insignificance of the variable. Model summary presenting value of R^2 is 97.67%, which is greater than the requisite criteria of 95%. Summary of this study is presented in Table 10.4.

TABLE 10.2
Experimental Results, GRC, and GRG

Sr No.	Response Variables			GRC Value			GRG Ideal 1	
	MT(AVG)	SR(AVG)	HRC(AVG)	HRC	SR	MT	GRG	RANK
1	60	0.754	42.1	0.5134	0.6829	0.6272	0.6078	8
2	57	1.438	38.4	0.4120	0.8930	0.5990	0.6347	6
3	60	1.562	50.5	1.0000	0.9526	0.6272	0.8599	2
4	68	1.803	41.1	0.4824	0.9910	0.7088	0.7274	4
5	90	1.934	45.3	0.6385	1.0000	1.0000	0.8795	1
6	50.4	0.214	41.3	0.4884	0.4509	0.5405	0.4933	13
7	57	0.212	38.3	0.4097	0.4509	0.5990	0.4865	14
8	74	1.524	48.2	0.8046	0.9550	0.7770	0.8455	3
9	34	0.324	41.1	0.4824	0.5090	0.4120	0.4678	16
10	37	1.384	43.9	0.5782	0.8779	0.4342	0.6301	7
11	64	0.278	41.2	0.4854	0.3544	0.6668	0.5022	11
12	31	0.218	38.3	0.4097	0.3387	0.3903	0.3796	25
13	33	0.961	47.4	0.7519	0.7554	0.4047	0.6373	5
14	34	0.263	40.1	0.4543	0.4846	0.4120	0.4503	17
15	38	0.257	43.8	0.5742	0.4679	0.4417	0.4946	12
16	30	0.198	38.7	0.4190	0.4419	0.3831	0.4147	21
17	32	0.187	40.4	0.4624	0.4265	0.3975	0.4288	19
18	35	0.172	41.6	0.4975	0.4225	0.4194	0.4464	18
19	43	0.217	40.2	0.4570	0.3333	0.4801	0.4235	20
20	24	0.334	43.4	0.5589	0.5258	0.3404	0.4750	15
21	32	0.179	35.9	0.3600	0.4439	0.3975	0.4005	23
22	41	0.312	34.4	0.3333	0.3682	0.4646	0.3887	24
23	23	0.483	47.3	0.7457	0.5873	0.3333	0.5554	10
24	24	0.187	39.6	0.4412	0.4519	0.3404	0.4112	22
25	24	1.212	43.2	0.5515	0.8265	0.3404	0.5728	9

TABLE 10.3
Response Table for Means

Level	P	AFR	SOD	Q
1	0.7419*	0.4837	0.5047	0.4971
2	0.5847	0.4637	0.5203	0.5456
3	0.4928	0.6689*	0.5470	0.5610
4	0.4377	0.4960	0.5880*	0.5111
5	0.4657	0.6104	0.5628	0.6079*
Max-min	0.3042	0.2053	0.0833	0.1107
Rank	1	2	4	3

* represents the highest values.

TABLE 10.4
Summary

Setting	Initial Setting Of Machine	Experimental Optimum Setting	Predicted Setting
Level	1-3-3-3	1-5-5-5	1-3-4-5
GRG value	0.8599	0.8795	0.973

10.4 CONCLUSIONS

The following conclusions are drawn from this study:

1. This study was focused on the analysis of experimental results using the GRA technique and results are very close to the ideal value (i.e., 1); the value is 0.8795 for optimal experimental setup and 0.973 for predicted setup.
2. Multiobjective optimum setup is 25 kpsi pressure, 300 g/min AFR, 10-mm SOD, and quality 5. Pressure is found to be the most impactful parameter, whereas SOD has the least impact on responses.
3. Increase in the pressure reduces the MT and SR. AFR plays a lead role in the improvement of hardness of the machined surface.

REFERENCES

Ducato A., Fratini L., La Cascia M., and Mazzola, G., 2013. An automated visual inspection system for the classification of the phases of Ti-6Al-4V titanium alloy, International Conference on Computer Analysis of Images and Patterns at Springer, Berlin, Heidelberg, pp. 362–369.

Hong S. Y., Markus I., and Jeong W. C., 2001. New cooling approach and tool life improvement in cryogenic machining of titanium alloy Ti-6Al-4V, *International Journal of Machine Tools and Manufacture* 41(15), 2245–226.

Joel C., and Jeyapoovan T., 2020. Optimization of machinability parameters in abrasive water jet machining of AA7075 using Grey-Taguchi method, *Materials Today: Proceedings* 37(2), 737–774.

Khan M.Y., and Rao P.S., 2019a. Electrical discharge machining: vital to manufacturing industries, *International Journal of Innovative Technology and Exploring Engineering* 8(11), 1696–1701.

Khan M.Y., and Rao P.S., 2019b. Optimization of process parameters of electrical discharge machining process for performance improvement, *International Journal of Innovative Technology and Exploring Engineering* 8(11), 3830–3836.

Khan M.Y., Rao P. S., and Pabla B.S., 2020. Investigations on the feasibility of jatropha curcas oil-based biodiesel for sustainable dielectric fluid in EDM process, *Materials Today: Proceedings* 26, 335–340.

Kumar S., Ghoshal S.K., and Arora P.K., 2021. Optimization of process variables in electric discharge machining (EDM) using Taguchi methodology, *Indian Journal of Engineering and Materials Sciences* 27(4), 819–825.

Kumar S., Ghoshal S.K., Arora P.K., and Nagdeve L., 2020. Multi-variable optimization in die-sinking EDM process of AISI420 stainless steel, *Materials and Manufacturing Process* 36(5), 572–582.

Lin Z.C., and Ho C.Y., 2003. Analysis and application of Grey relation and ANOVA in chemical–mechanical polishing process parameters, *International Journal of Advanced Manufacturing Technology* 21, 10–14.

Momber A.W., and Kovacevic R., 2012. *Principles of Abrasive Water Jet Machining*, Springer Science & Business Media, Cham, Switzerland.

Nair A., and Kumanan S., 2018. Optimization of size and form characteristics using multiobjective Grey analysis in abrasive water jet drilling of Inconel 617, *Journal of the Brazilian Society of Mechanical Sciences and Engineering* 40(3), 1–15.

Naresh Babu M., and Muthukrishnan N., 2015. Investigation of multiple process parameters in abrasive water jet machining of tiles, *Journal of the Chinese Institute of Engineers* 38(6), 692–700.

Rao P.S., Jain P.K., and Dwivedi D.K., 2014a. Effect of electrolyte composition on surface finish of titanium alloy by electro chemical honing process, 1st International conference on ICNDME- 2014 at MMU, Mullana (Ambala).

Rao P.S., Jain P.K., and Dwivedi D.K., 2014b. Study and prediction of surface finish of external cylindrical surfaces of titanium alloy by electro chemical honing (ECH), 23rd International Conference on PFAM at IITR, Roorkee, India.

Rao P.S., Jain P.K., and Dwivedi D.K., 2015a. Electro chemical honing (ECH) of external cylindrical surfaces of titanium alloy, *Procedia Engineering* 100, 936–945.

Rao P.S., Jain P.K., and Dwivedi D.K., 2015b. Study and effect of process parameters on external cylindrical surfaces of titanium alloy by electro chemical honing (ECH), 26th DAAAM International Symposium on Intelligent Manufacturing and Automation, Zadar, Croatia.

Rao P.S., Jain P.K., and Dwivedi D.K., 2016a. Optimization of key process parameters on electrochemical honing (ECH) of external cylindrical surfaces of titanium alloy Ti-6Al-4V, 25th International Conference on Materials Processing and Characterization ICMPC at GRIET, Hyderabad, India.

Rao P.S., Jain P.K., and Dwivedi D.K., 2016b. Prediction of optimal process parameters on to electro chemical honing (ECH) of external cylindrical surfaces of titanium alloy using DOE technique, 6th International and 27th All India Manufacturing Technology, Design and Research Conference (AIMTDR 2016) at COEP, Pune, India.

Rao P.S., Jain P.K., and Dwivedi D.K., 2017. Optimization of key process parameters on electro chemical honing (ECH) of external cylindrical surfaces of titanium alloy Ti-6Al-4V, *Materials Today Proceedings* 4(2), Part A (2017), 2279–2289.

Ross P.J., 1996. *Taguchi Techniques for Quality Engineering*, 2nd ed. McGraw-Hill, New York.

Roy R.K., 2001. *Design of Experiments Using the Taguchi Approach: 16 Steps to Product and Process Improvement*, John Wiley & Sons, New York.

Senthilkumar T.S., Muralikannan R., and Kumar S. S., 2020. Surface morphology and parametric optimization of AWJM parameters using GRA on aluminum HMMC, *Materials Today: Proceedings* 22, 410–415.

Seo Y.W., Ramulu M., and Kim D., 2003. Machinability of titanium alloy (Ti'6Al'4V) by abrasive waterjets. *Proceedings of the Institution of Mechanical Engineers, Part B: Journal of Engineering Manufacture* 217(12), 1709–1721.

11 Electrochemical Machining of Cu Substrate with Cu Tool
A Case Study

Gurwinder Singh, Rupinder Singh, and P. Sudhakar Rao

CONTENTS

11.1 INTRODUCTION: BACKGROUND AND DRIVING FORCES

In conventional machining (CM), it is very necessary to get the desired accuracy on the material surface to avoid post-processing. Usually in CM the material is removed in the form of chips, flakes, and burrs. The desired surface features such as dimensional accuracy can be attained easily by nontraditional machining (NTM) processes such as electrochemical machining (ECM). ECM is opposite to the galvanic/electrochemical coating method or deposition process used to machine enormously hard materials that are very difficult to machine effectively using conventional methods (Westley et al., 2004). ECM works on Faraday's principle, which states that if two conductive poles are placed in a conductive electrolyte bath and energized by a current, metal may be de-plated from the positive pole (the anode) and plated onto the negative pole (the cathode). Thus, ECM can be used to remove the material from an electrically conductive workpiece through anodic dissolution (Bhattacharyya et al., 2002). It has been widely reported that mechanical or thermal energy is not involved in ECM. This process is usually widely used for mass production and has good applications for machining of extremely hard materials that are difficult to machine using CM. ECM has been extensively used in the production of semiconductor equipment and in aerospace and electronics

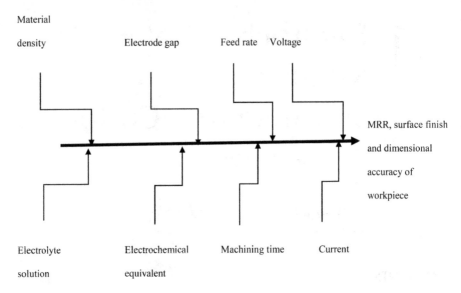

FIGURE 11.1   Cause-and-effect diagram for ECM.

industries (Datta and Romankiw, 1989). Based on reported literature, Figure 11.1 shows the cause-and-effect diagram for ECM.

To ascertain the research gap, bibliographic analysis was performed for the past 20 years using the keyword electrochemical machining on science databases. Results were obtained from science core collection 3912 on the Web. Out of these results, the top 500 cited articles were sorted and processed with viewer open source (VOS) software. Figure 11.2 shows a networking diagram based on bibliographic analysis (using the keyword electrochemical machining). Another search was made for the keyword electrochemical machining of Cu and 217 results were obtained. The corresponding networking diagram is shown in Figure 11.3. Finally another search was made using the keyword electrochemical machining of Cu with Cu tool and 46 results were obtained. In this case by selecting the minimum number of occurrences of terms as 5, a total of 48 terms met the threshold out of 1562 terms. For each 48 terms a relevance

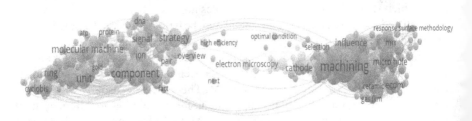

FIGURE 11.2   Networking diagram based on bibliographic analysis (for the keyword electrochemical machining).

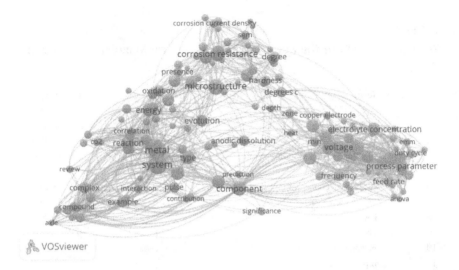

FIGURE 11.3 Networking diagram based on bibliographic analysis (for the keyword electrochemical machining of Cu).

score was calculated. Based on this score the top 60% terms (i.e., 29) were selected for analysis (Table 11.1). Further, based on Figure 11.4, Figure 11.5 shows research gap analysis (using the keyword electrochemical machining of Cu). The ECM setup is shown in Figure 11.6.

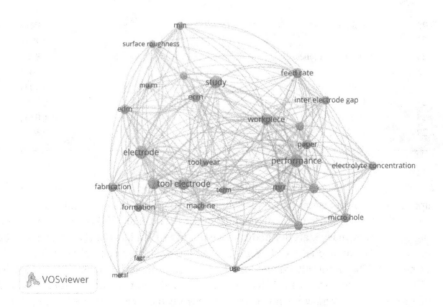

FIGURE 11.4 Networking diagram based on bibliographic analysis (for the keyword electrochemical machining of Cu with Cu tool).

TABLE 11.1

Relevance Score (for the Keyword Electrochemical Machining of Cu with Cu Tool)

Id	Term	Occurrences	Relevance Score
1	Application	13	0.5508
2	ECM	9	0.3426
3	EDM	7	1.2119
4	Electrochemical micromachining	7	0.4988
5	Electrode	17	0.725
6	Electrolyte concentration	6	1.7833
7	Fabrication	8	1.1501
8	Fact	5	2.8061
9	Feed rate	8	0.9676
10	Formation	8	1.5112
11	Input parameter	7	1.2176
12	Inter electrode gap	6	1.2854
13	Machine	9	0.4175
14	Material removal rate	7	0.5476
15	Metal	6	4.0026
16	Micro hole	7	1.0604
17	Min	5	0.6423
18	MRR	7	0.4746
19	Mu m	5	0.8596
20	Paper	9	0.9999
21	Performance	14	0.4355
22	Stainless steel	6	0.7412
23	Study	16	0.4941
24	Surface roughness	6	0.898
25	Term	5	0.7058
26	Tool electrode	12	0.3031
27	Tool wear	6	0.7352
28	Use	6	1.0262
29	Workpiece	11	0.6059

The literature review reveals that due to electrical and thermal properties, Cu has applications in precision instruments in the widespread areas of automobile, aerospace, electronics, instrumentation, and defense (Tzou et al., 2004). Precise machining for extraordinary stability and surface quality of Cu has a direct impact on the device's performance (Purkait et al., 2018), thus requiring special attention while using CM. The main advantages of ECM are high material removal rate (MRR) of the workpiece, almost no MRR of the tool, and good surface texture deprived of re-solidified layers with no residual stress on the machined part (Jain, 2008; Klocke et al. 2012; Liu and Qu 2019). Feng et al. (2020) outlined the nozzle dwell time and path control mechanism for obtaining a complex three-dimensional (3D)

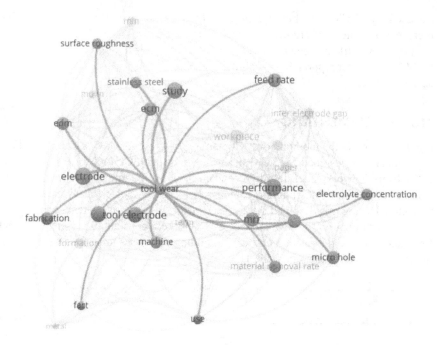

FIGURE 11.5 Research gap based on bibliographic analysis (for the keyword electrochemical machining of Cu with Cu tool).

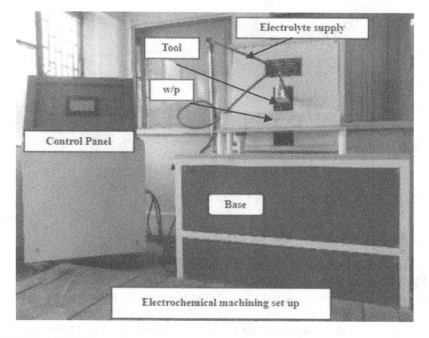

FIGURE 11.6 ECM setup.

surface by jet ECM. The bibliographic analysis for the past 20 years outlined that there are a number of studies on machining of Cu substrate with different tool materials, but little has been reported on the use of the Cu tool on Cu substrate by ECM. This chapter reports the machining capabilities of the Cu tool with Cu substrate as a case study.

11.2   EXPERIMENTATION

Figure 11.7 shows the methodology adopted for the present study. A case study base experiment was designed for ECM of Cu substrate with Cu tool for observing the MRR from the workpiece, stability of the electrolyte, and rate of tool wear by controlling various input machining parameters (feed rate of tool, electrolyte concentration, etc.). During this designed set of experiments, the current, voltage, and desired time of machining and tool-workpiece gap were kept constant. The selected input fixed and variable parameters and corresponding responses at different settings in the experiment are shown in Table 11.2.

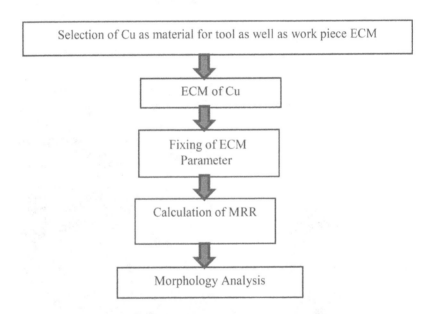

FIGURE 11.7   Methodology adopted for present study.

11.3   RESULT AND DISCUSSION

Based on the observed values from the Table 11.2, to get the best results in terms of MRR for a tool as well as workpiece material by using response surface methodology was used to compare individual and group response on input setting parameters to get the optimized result. The contour plot of tool wear and MRR of a work-piece is well compared by means of contour plot by using response surface modeling (RSM) in Figure 11.8(a) and (b).

TABLE 11.2

Machining Parameters Considered During ECM of Cu

S. No.	Feed Rate (μ/min) (A)	Concentration (g/L) (B)	Voltage (V)	Time (s)	Current (I)	Workpiece Weight Before ECM (g)	Workpiece Weight After ECM (g)	Tool Weight Before ECM (g)	Tool Weight After ECM (g)	Electrode Gap (mm)	MRR Workpiece (mg/min) (C)	MRR Tool (mg/min) (D)
1	144	100/1	25	1200	300	210.2152	209.8217	49.4324	49.4178	2.376	2.0056	0.0746
2	180	100/1	25	1200	300	209.8217	207.8690	49.4178	49.4002	2.376	9.9525	0.0897
3	216	100/1	25	1200	300	205.0063	202.2341	49.4002	49.3964	2.376	14.1438	0.0193
4	144	150/1	25	1200	300	202.2341	195.2267	49.3964	49.3862	2.376	35.7156	0.0519
5	180	150/1	25	1200	300	195.2267	192.9295	49.3862	49.3786	2.376	11.7084	0.0387
6	216	150/1	25	1200	300	192.9295	189.7090	49.3786	49.3621	2.376	16.4143	0.0841
7	144	200/1	25	1200	300	189.7090	188.6501	49.3621	49.3545	2.376	5.3970	0.3873
8	180	200/1	25	1200	300	188.6501	185.2410	49.3545	49.3412	2.376	17.3756	0.6778
9	216	200/1	25	1200	300	185.2410	182.9991	48.3412	48.3217	2.376	11.4266	0.9938

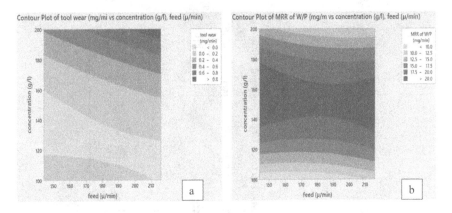

FIGURE 11.8 (a) Contour plot for wear versus concentration and feed rate and (b) Contour plot of MRR of workpiece versus concentration and feed rate.

Finally, based on the observations of MRR of the workpiece and tool wear (Table 11.2), multiresponse optimization was performed by using Minitab 19 by giving equal weight to both MRR of the workpiece and tool for comparative result analysis (Figure 11.9). The optimal composite desirability was 0.7710 at optimized setting for both cases at a feed rate of 216 µ/min and rate of concentration of 200 g/L.

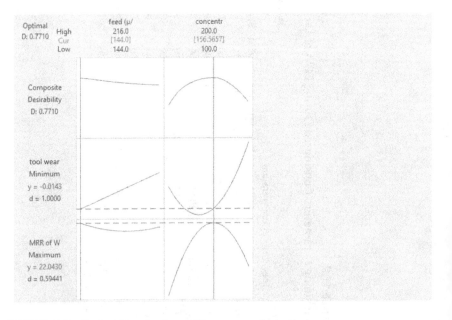

FIGURE 11.9 Desirability plot for multiresponse optimization.

11.4 CONCLUSIONS

The following are the conclusions from this study:

- ECM of Cu workpiece by Cu tool was gainfully adopted with controlled input parameters to obtain a higher MRR from the workpiece with a small expense of tool wear.
- From the multifactor optimization viewpoint better MRR of the workpiece was achieved at a feed rate of 18 μ/min and concentration of 200 g/L.

Further studies may be performed with different electrolyte by changing the tool dimensions.

ACKNOWLEDGMENTS

The authors thank NITTTR, Chandigarh, and Panjab University, Chandigarh, for use of their lab facilities.

REFERENCES

Bhattacharyya B., Mitra S., and Boro A.K., 2002. Electrochemical machining: new possibilities for micromachining, *Robotics and Computer-Integrated Manufacturing* 18(3–4), 283–289.

Datta M., and Romankiw L.T., 1989. Application of chemical and electrochemical micromachining in the electronics industry, *Journal of the Electrochemical Society* 136(6), 285C.

Feng Y., Yan Y., Zhou P., and Guo D., 2020. High precision material removal of copper surface by jet electrochemical machining, *IOP Conference Series: Materials Science and Engineering* 715, 012052.

Jain V.K., 2008. Abrasive-based nano-finishing techniques: an overview, *Machining Science and Technology* 12(3), 257–294.

Klocke F., Zeis M., Klink A., and Veselovac D., 2012. Technological and economical comparison of roughing strategies via milling, EDM and ECM for titanium-and nickel-based blisks. *Procedia CIRP*, 2, 98–101.

Liu Y., and Qu N., 2019. Electrochemical milling of TB6 titanium alloy in $NaNO_3$ solution, *Journal of the Electrochemical Society* 166(2), E35.

Purkait T., Singh G., Kumar D., Singh M., and Dey R.S., 2018. High-performance flexible supercapacitors based on electrochemically tailored three-dimensional reduced graphene oxide networks, *Scientific Reports* 8(1), 1–13.

Tzou H.S., Lee H.J., and Arnold S.M., 2004. Smart materials, precision sensors/actuators, smart structures, and structronic systems, *Mechanics of Advanced Materials and Structures* 11(4–5), 367–393.

Westley JA., Atkinson J., and Duffield A., 2004. Generic aspects of tool design for electrochemical machining, *Journal of Materials Processing Technology* 149(1–3), 384–392.

Section III

Smart Manufacturing

12 A Survey on Existing Facilities and Infrastructure in Small-Scale Machining Industries
Dimensions and Future Prospects

Mohd Bilal Naim Shaikh and Mohammed Ali

CONTENTS

12.1 INTRODUCTION

Machining encompasses a broad range of technologies as well techniques through which excess material from a workpiece is removed in the form of chips to shape it into an intended design. In a typical machining process, the workpiece undergoes shearing, bending, and compression phenomena due to the action of the cutting tool (Childs et al., 2000; Arrazola et al., 2013). The orthogonal machining

DOI: 10.1201/9781003203681-15

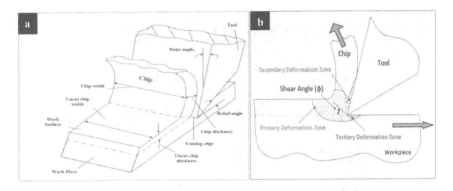

FIGURE 12.1 Schematic illustration of the (a) Machining process and (b) Different heat generation sources.

process is considered to be the most straightforward geometrical representation of the machining process (Figure 12.1(a)). In it, the cutting tool edge remains normal to the direction of relative motion between tool and workpiece, and the generated new surface plane used to be parallel to the original one of the work materials (Merchant, 1945a,b). In practice, for an excellent orthogonality approximation of a machining process, it must possess the following characteristics: depth of cut quite large compared with feed, the radius of the workpiece should be large compared with the depth of the cut, and there should be a few degrees of obliquities of the cutting edge and insignificant tool radius (Waldorf et al., 1998). All the phenomena happening during the machining process rotate around the three tool-workpiece-chip interactions: shear zone (i.e., primary deformation zone), chip-tool friction zone, (i.e., secondary deformation zone), and the finished workpiece-tool interface (i.e., tertiary deformation zone; Figure 12.1(b)) (Fang, 2003; Priyadarshini et al., 2012; Uysal and Altan, 2016). Heat generation due to plastic deformation and friction along with the combined loading effects of the tool has a very significant impact on the fundamental variables such as cutting temperature, stress-strain, microstructures and strain rates, temperatures, etc. These variables are greatly correlated to the process performance measures like burr formation, chip form/breakability, cutting temperature, surface integrity, tool life, and others.

Generally, the machining industries transform raw materials into the finished workpiece. This transformation has different resources (materials, energy, cutting fluids, technical knowledge, etc.) at the input side and different entities (waste streams, entropy, finished product, emissions, etc.) on the output end. The relations between these inputs, outputs, and machine tools during the metal removal process are illustrated in Figure 12.2. Principal activities like production of materials, tool preparation, cutting fluid preparation, removal of excess material, and cleaning of the finished product are significant contributors to the overall machining scenario. From the system analysis, it can be concluded that optimal material usage, minimal cutting fluid application, and minimal power consumption (i.e., cutting energy, reduce emissions, and minimal environmental and health hazards)

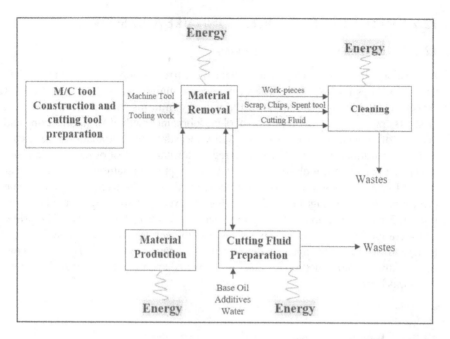

FIGURE 12.2 Relations of machinery and equipment, as well as material and energy, that flow within a machining industry.

are the themes on which researchers have conducted a great deal of research. All of these process level activities and their improvements have crucial environmental ramifications.

Aligarh is known worldwide for its lock and hardware industries. In 1890, local production of locks and their parts were initiated by small entrepreneurs here on a small scale. At present, the city holds thousands of manufacturers, exporters, and suppliers that are associated with zinc die, metal-based (steel and aluminum fab) electrical machinery and transport equipment, engineering and workshop units, and automobile and defense components-based industries. Machining industries are under pressure to cope with competition, environmental regulations, and supply chain demand for better working environment conditions (Rosen and Kishawy, 2012; Kishawy et al., 2018). Hence, it is necessary for all companies to adopt sustainable practices for improved economic, environmental, and social performance. Basic sustainability principles are concerned with manufacturing costs, power consumption, waste management, environmental impact, operational safety, and personal health.

This chapter seeks to find out the existing status of machining industries located within Aligarh. A survey has been performed to find out the nature of work, process, products, and the issues associated with the current facilities in their organizations. Further, the future initiatives and advancements that are needed for improving the overall performance of the firm have also been identified. Based on the observations, the solution to each barrier and the need for implementation of sustainable based practices, resources, and methodologies are also discussed.

12.2 METHODOLOGY: QUESTIONNAIRE AND SURVEY

12.2.1 DEVELOPMENT OF QUESTIONNAIRE

A questionnaire was developed to collect the current status of machining industries/workshops for this study. The questionnaire mainly focused on four different aspects: (1) the basic information of the industry in terms of workpiece materials, products, process etc.; (2) types of coolant/lubricant, method of application, and quantity of coolant/lubricant; (3) the quality measures and automation levels, etc.; and (4) innovation/initiatives for improving performance. Most of the questions are structured ones that are closed ended and multi-option in nature. Further, innovation and initiative-based questions were developed using the five-point Likert scale principle. The scale ranges from 1 to 5 where 1 = very low impact, 2 = medium low impact, 3 = average impact, 4 = high impact, and 5 = very high impact. The prime consideration in the design of this survey was to keep it short (i.e., about 20 questions) and focused to receive specific as well as adequate responses. This survey did not use any sampling method to select participants, and the total number of respondents was kept at 50.

12.2.2 INDUSTRIAL SURVEY

The survey was conducted in the city of Aligarh, which is famous for its hardware industry, to see the current scenario of the machining process in various industries. The survey was done through personal interviews with manufacturing and/or production managers, engineers, trainees, quality control and/or assurance personnel, and technicians. During the survey, there were special concerns regarding the various components associated with the machining process. Sustainability concerns in machining industries were identified and alternatives for them were suggested.

12.3 FINDINGS AND ANALYSIS

12.3.1 PROCESSES, RAW MATERIALS, AND PRODUCTS

The survey concluded that the majority of the industries were employing conventional-based machine tools and processes (Figure 12.3(a)), which could be from their small-scale units, orders of products, and the financial condition of the firms. Around 61% of total responses employed conventional machining processes, whereas about 25% of the total were employing nonconventional ones. Further, some fraction is involved in secondary machining processes, which is needed for the semi-finished products. As far as material of workpiece is concerned, various types of steel like mild steel, alloy steel, and tool steel are used (Figure 12.3(b)). The concerned cutting tool materials used are shown in Figure 12.3(c). Further, the product pie chart (Figure 12.3(d)) shows that machine parts, automotive components, die-punch assembly, and lock ware are the major products.

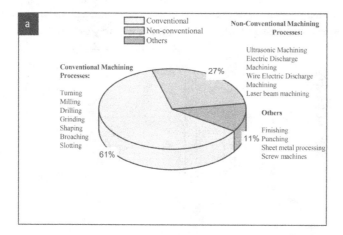

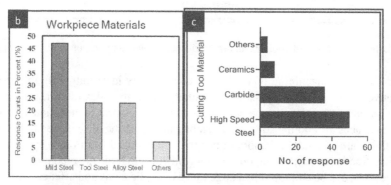

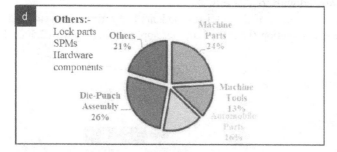

FIGURE 12.3 Respondents distribution regarding (a) Types of machining, (b) Workpiece material used, (c) Tool material, and (d) Products developed.
Abbreviations: SPM, special purpose machines.

12.3.2   COOLANT/LUBRICANT AND THEIR APPLICATION METHODS

It was found from the survey (Figure 12.4) that the following categories are in use by the respondents: neat oil or straight oils (mineral based), soluble oils or emulsions, synthetic or chemical cutting fluids, semi-synthetic cutting fluids, and plain water. Mineral-based oils are in two basic categories, namely, paraffinic and naphthenic.

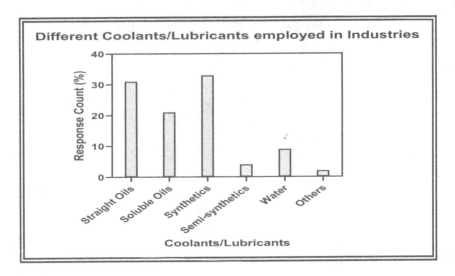

FIGURE 12.4   Types of coolants and/or lubricants employed by respondents.

These types of machining fluids usually contain fatty lubricants, extreme pressure additives, polar additives, thickness modifiers, viscosity index modifiers, and odorants for characteristic enhancement. Soluble oils are a kind of emulsion with water that enhances the cooling ability of the lubricant. Synthetic oils referred to as chemical fluids are generally alkanol amines that are deduced from other hydrocarbons and by fractionation of crude oils. Semi-synthetics, hybrids of soluble oils and synthetics oils, can be employed in a large range of applications and offer easier maintenance compared with soluble oils.

Further, two-thirds of the total respondents utilized the conventional method of cutting fluid application (i.e., flood and jet application) (Figure 12.5(a)). Because of

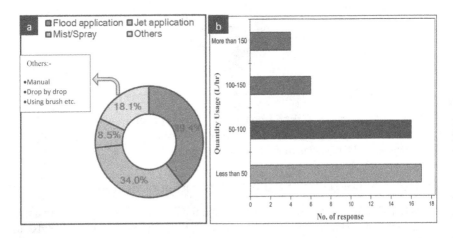

FIGURE 12.5   (a) Method of application used for coolant/lubricants and (b) Quantity utilized during the machining process.

this an extravagant quantity of machining fluid (Figure 12.5(b)) needs to be stored and enough fraction waste is generated due to splitting during application. In addition, several issues were associated with the usage of these conventional cutting fluids and their application methods (Figure 12.6).

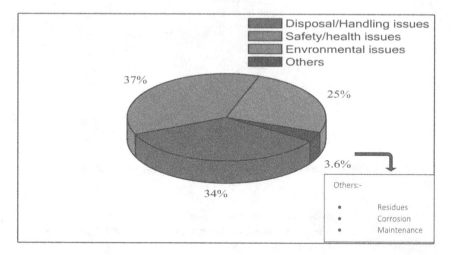

FIGURE 12.6 Problems/issues associated with the application of conventional cutting fluids.

12.3.3 AUTOMATION AND QUALITY CONTROL AND/OR MANAGEMENT

To meet the demand and quality as per requirement, automation and numerical control-based practices are needed. Hence, respondents were asked to categorize the level of automation utilized on their shop floor, and observations are presented in Figure 12.7(a). Furthermore, products from the machining process yield distribution of quality due to variation of size, dimension, geometrical tolerance, and geometrical

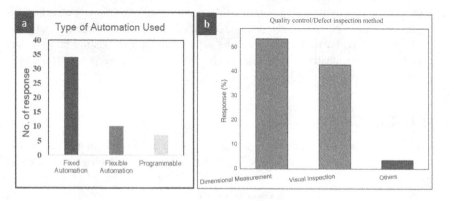

FIGURE 12.7 (a) Automation level used and (b) Quality control methods employed by respondents.

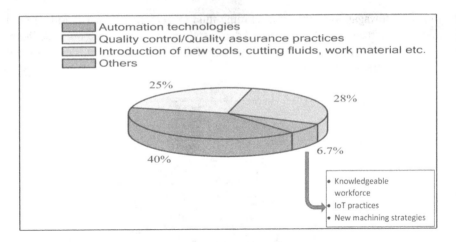

FIGURE 12.8   Innovations and/or investments to be introduced by respondents in the coming years.

properties of the surface produced. Because of this, organizations set up quality control units and use various measurement methods to extract the defective products. The same for surveyed industries is shown in Figure 12.7(b).

12.3.4   FUTURE INITIATIVES AND THEIR IMPACT ON PERFORMANCE

In the last section of the survey, respondents were asked about the change in their existing status to improve the functioning and overall productivity of the organization. Closed-end questions with choices of new quality control and/or assurance practices, automation methods, and introduction of new machining parametric conditions were provided to respondents. Most people are ready for advancing existing facilities with some level of automation (i.e., numeric control), whereas around half of the respondents chose to improve quality control methods and new materials (tool, workpiece, cutting fluids, etc.), respectively (Figure 12.8). Some respondents were willing to employ Internet of Things (IoT) concepts, nonconventional machine tools, skilled workforce, etc. Similarly, respondents rated the impact of these new advancements on a scale of 1 to 5 units. It can be concluded that the existing facilities and infrastructure of industry need to improve because none of the respondents are satisfied with the same (Table 12.1).

12.4   DISCUSSION

Fundamental observations and the associated issues/problems mentioned earlier need to be discussed for the best possible solution. Sustainable product design and manufacturing principles are being used nowadays for the improvement of all of the presented levels. Small-scale machining companies are traditionally centered on short-term financial considerations, which in turn adversely affect the longer term view. Hence, adaptation of sustainable manufacturing practices results in a more prosperous,

TABLE 12.1
Predicted Impact of New Practices/Materials on Overall Performance of a Firm

Impact of Innovations on Industry's Overall Performance Implementation/Introduction of	Very Low (1)	Medium Low (2)	Average (3)	Low High (4)	Very High (5)
New workpiece materials	–	5	25	19	2
New cutting tool/inserts	–	12	21	16	2
New cutting fluids and application methods	–	14	18	15	3
Automation technologies	–	2	20	6	22

cleaner, safer, and healthier environment and lower operating costs, which increase the bottom line profits, competitiveness, public profile/image for the company, and create a fully prepared system for future variations with regard to health, safety, and environmental legislation. The following are some suggestions with respect to the previous observations from the survey (Haapala et al., 2013; Debnath et al., 2014; Ghandehariun et al., 2016; Sharif et al., 2017; Krolczyk et al., 2019; Lee et al., 2019).

12.4.1 EFFICIENT USE OF ADVANCE RESOURCES AND WASTE MINIMIZATION

The main resources concerning the machining process are as follows:

- *Workpiece material:* Key features for a sustainable alternative include abundant supply, potential for recycling/reuse, and minimal environmental impact.
- *Coolants/lubricants/hydraulic oils:* These are less toxic, sustainable, and environment- and operator-friendly.
- *Energy:* Optimization of machining parameters, regular inspection, and maintenance of machine tools, and zero waste production.
- *Waste minimization:* The hierarchy in Figure 12.9 has to be followed for systematic reduction of waste.

FIGURE 12.9   Waste reduction approach.

12.4.2 PRODUCTIVITY

For optimization of process parameters, in general higher cutting parameters do result in a higher rate of material removal (i.e., short processing time), but it occurs at the expense of a reduced tool life, which adversely affects the productivity and efficiency.

For process automation, monitoring and control of the machine process yields an increment in productivity, product quality, reduced costs, and relaxed machine design constraints. Surface integrity, dimensional accuracy, tool condition, cutting temperature, machining force, and chatter detection are some of the key parameters to be controlled and monitored regularly for better performance.

12.4.3 ADVANCE TOOL MATERIALS AND COATINGS

Tool material and its geometry play an important role in the machining performance as the cutting tool operating life follows certain wear criterion principles. A worn out cutting edge is not only considered to be unstable, but it also adversely affects the dimensional accuracy and surface integrity of the machined surface. Hence, new materials like coated cemented carbides, polycrystalline diamond, and polycrystalline boron nitride are being employed. Further, coatings of TiN, TiCN, and Al_2O_3 are also being widely used to enhance tool life.

12.4.4 HYBRID MACHINING PROCESSES

Hybrid machining processes are a combination of nonconventional manufacturing techniques and mechanical (conventional) machining processes to improve the machinability performance. Introduction of hybrid machining not only reduces the tool failure frequency but also improves the work surface integrity at the expense of machining cost. Laser- and ultrasonic-assisted machining processes are generally utilized when there are difficult-to-cut-materials.

12.4.5 GREEN CUTTING FLUIDS AND APPLICATION TECHNIQUES

Cutting fluid controls the thermal and tribological variations in the workpiece and cutting tool during the process. However, conventional cutting fluids and application methods significantly affect the operator's health and contribute to the overall cost as well as environmental pollution. Vegetable oils, esters, and gas-based coolants are a few sustainable alternatives, whereas the near-dry cooling method is a more economic and less polluting method of application.

12.5 CONCLUSIONS

Based on the observations of the industrial survey and the earlier discussion, the following conclusions and recommendations can be made:

1. The survey, conducted in the metal machining industry, used a total of 50 responses for assessing their current status in terms of type of process, workpiece material, cutting fluids, etc.
2. It is observed that the conventional machining process, coolant/lubricant, and cutting tool material are in use by the majority of respondents.
3. Fixed automation and manual inspection of products are the common practices/techniques employed by the respondents.

4. Sustainability principles are to be incorporated for better overall performance of the machining industry, which not only enhances the overall profit, competitiveness, and public profile/image of the industry, but also keeps it up to date for future health, safety, and environmental legislation.

REFERENCES

Arrazola P.J. et al., 2013. Recent advances in modelling of metal machining processes, *CIRP Annals - Manufacturing Technology* 62(2), 695–718.

Childs T.H.C., Maekawa K., Obikawa T., and Yamane Y., 2000. *Metal Machining - Theory and Applications*. Elsevier, Amsterdam.

Debnath S., Reddy M.M., and Yi, Q.S., 2014. Environmental friendly cutting fluids and cooling techniques in machining: a review, *Journal of Cleaner Production* 83, 33–47.

Fang N., 2003. Slip-line modeling of machining with a rounded-edge tool - Part I: new model and theory, *Journal of the Mechanics and Physics of Solids* 51(4), 715–742.

Ghandehariun A., Nazzal Y., and Kishawy H., 2016. Sustainable manufacturing and its application in machining processes: a review, *International Journal of Global Warming* 9(2), 198–228.

Haapala K.R. et al., 2013. A review of engineering research in sustainable manufacturing, *Journal of Manufacturing Science and Engineering* 135(4), 041013.

Kishawy H.A., Hegab H., and Saad, E., 2018. Design for sustainable manufacturing: approach, implementation, and assessment, *Sustainability* 10(10), 1–15.

Krolczyk G.M. et al., 2019. Ecological trends in machining as a key factor in sustainable production – a review, *Journal of Cleaner Production* 218, 601–615.

Lee H.T. et al., 2019. Research trends in sustainable manufacturing: a review and future perspective based on research databases, *International Journal of Precision Engineering and Manufacturing-Green Technology* 6(4), 809–819.

Merchant M.E., 1945a. Mechanics of the metal cutting process. I. orthogonal cutting and a type 2 chip, *Journal of Applied Physics* 16(5), 267–275.

Merchant M.E., 1945b. Mechanics of the metal cutting process. II. Plasticity conditions in orthogonal cutting, *Journal of Applied Physics* 16(6), 318–324.

Priyadarshini A., Pal S.K., and Samantaray A.K., 2012. Finite element modeling of chip formation in orthogonal machining. In Davim, J.P. ed., *Statistical and Computational Techniques in Manufacturing*. Springer, Berlin, pp. 101–144.

Rosen M.A., and Kishawy, H.A., 2012. Sustainable manufacturing and design: concepts, practices and needs, *Sustainability* 4(2), 154–174.

Sharif M.N., Pervaiz S., and Deiab I., 2017. Potential of alternative lubrication strategies for metal cutting processes: a review, *The International Journal of Advanced Manufacturing Technology* 89, 2447–2479.

Uysal A., and Altan, E., 2016. Slip-line field modelling of rounded-edge cutting tool for orthogonal machining, *Proceedings of the Institution of Mechanical Engineers, Part B: Journal of Engineering Manufacture* 230(10), 1925–1941.

Waldorf D.J., DeVor R.E., and Kapoor S.G., 1998. A slip-line field for ploughing during orthogonal cutting, *Journal of Manufacturing Science and Engineering* 120(4), 693–699.

13 An Overview of Current Market Conditions of Electric Vehicles

Manav Singh and Sandeep Grover

CONTENTS

13.1 INTRODUCTION: BACKGROUND AND DRIVING FORCES

Since the late 19th-century pre-industrial period (1850s to the present time), the Earth has experienced a massive change in its environment due to industrialization, and the Earth's surface temperature has risen higher and higher. Overexploitation of fossil fuels and uses of such conventional sources of energy for transportation needs do bring change, but industrialization comes at the cost of more greenhouse gas emissions and a further rise in the Earth's temperature.

As we limit the use of fossil fuels as our primary source of energy and change from conventional to nonconventional sources of energy for industrial and daily use, we also change our methods for transportation. One alternative is electric vehicles (EVs), as they are more efficient than conventional vehicles and produce no direct emissions. The automotive industry has accepted this and is moving toward EVs as an alternative. Thus, the State of India has different prospects due to the challenges of India's infrastructure.

EVs have huge potential as a better solution for future transportation needs until and unless more efficient solutions are found for the challenges that current EV technology is facing. These challenges include range concern; less frequent charging of batteries; zero pollution for the current state of lithium-ion battery manufacturing and its disposal waste, which is still harmful to the environment; and charging infrastructure and its maintenance. Also, complete battery-powered EVs are dependent on lithium ion as a source, which is still a limited natural resource. An alternate solution needs to be found such as ultra-capacitors or fuel cell technology, which

DOI: 10.1201/9781003203681-16

are in the development stages. The working of EVs is also very simple, as shown in Figure 13.1.

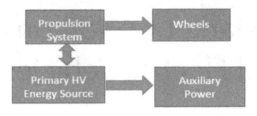

FIGURE 13.1   Pure electric vehicle framework.

Norway has the highest number of EV consumers, and other countries like the United States and China, which have a similar economy to India, have managed to build a concrete plan for the EV market of the future, including building the substantial infrastructure required for the success of EVs. China is the largest manufacturer of EVs because it is the fourth largest source of lithium ion, making it a leader in battery manufacturing technology. This has helped China to build its own market for EVs along with government policies mandating the construction of charging stations. China holds 3.3% of the consumer market for EVs, whereas India holds only 0.1%. Figure 13.2 illustrates the plug-in electric car ownership per capita in selected top-selling countries and regional markets.

India's fast-growing economy has presented huge opportunities for the developing industrialist to set up future electric mobility. India is in the initial development stage

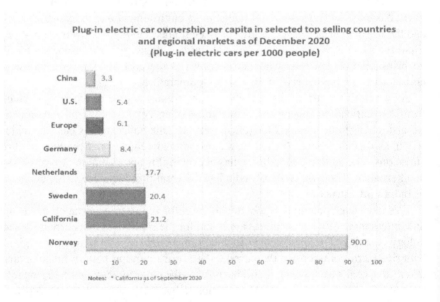

FIGURE 13.2   Comparison of plug-in electric car ownership per capita in selected top-selling countries and regional markets as plug-in cars per 1000 people (2020). (From Mariordo (Mario Roberto Durán Ortiz), 2020.)

for electric mobility and is set for a better future with EVs as necessary technological advancement, automation, and robust and flexible designs are becoming available. This also includes huge manpower potential and innovation in India, but there are presently challenges regarding demand of Indian consumers and limited access to infrastructure.

13.2 ELECTRIC VEHICLES

13.2.1 CURRENT SCENARIO

In the current scenario of the Indian EV market, there are many challenges that consumers and suppliers are facing in establishing EVs in the Indian market.

The following are challenges that consumers face:

1. Concrete EV four-wheel designs are always made for better efficiency to provide more range and are built with standards for good road conditions and to reduce drag. Yet, Indian road conditions are not always good; rather, the driving style of Indian consumers and improper road standards affect and reduce the efficiency of EVs in city drive conditions.

2. Design plays a major role in EV efficiency; hence, EVs with a better drag coefficient have lower ground clearance, which again is affected by improper road standards and speed breakers. Designs also affect the efficiency and prospect of Indian consumers.

3. Also, design parameters are based on different vehicles and can vary from one original equipment manufacturer (OEM) to other, despite the design prospect of the vehicle. Consumers are still concerned about range issues and worry about finding a charging station nearby when driving for long drives. Charger installation is to be provided based on statewide polices. Construction is underway and has gained momentum after the government allotted INR 1000 crore (approximately $135 million) for charging infrastructure under the FAME II policy. The Department of Heavy Industries (DHI) sanctioned 2636 new charging stations across 62 cities and 24 states and union territories in January 2020. Under this scheme at least one charging station will be set up every 3 km. Hence, consumers need to be educated accordingly to help them understand that EV infrastructure is being built and it is feasible with current OEMs. The government can emphasize education with strong rules that will further incentivize consumers such as subsidy schemes that will benefit consumers in multiple states, including 12 states that will support infrastructure and 11 of those will also provide EV incentive schemes to benefit consumers. Road tax waivers in all states will help build the ecosystem. The government is further emphasizing EVs and their manufacture. Figure 13.3 shows Indian state-wise incentive schemes.

4. India is at the developing stage, and the fourth major concern of consumers is cost-effectiveness as they think that the initial cost of EVs is very high and the technology has a higher initial cost and range limitation. Because of this, it was difficult to convince states to move toward using EVs. But if we

States	Targets	Financial incentives	Infrastructural incentives
Andhra Pradesh	Yes	Yes	Yes
Bihar	Yes	Yes	Yes
Delhi	Yes	Yes	Yes
Karnataka	Yes	Yes	Yes
Kerala	Yes	Yes	No
Madhya Pradesh	No	Yes	Yes
Maharashtra	No	Yes	Yes
Punjab	No	Yes	Yes
Tamil Nadu	Yes	Yes	Yes
Telangana	Yes	Yes	Yes
Uttar Pradesh	Yes	Yes	Yes
Uttarakhand	No	No	No

FIGURE 13.3 Indian state-wise incentive schemes released by the substates. (From NITI-Aayog and World Council, 2018.)

see a complete picture of the average driving time of the Indian consumer who has an individual vehicle, then after complete summation of fuel cost, running cost, maintenance cost, and overall cost of vehicle, we find that EVs are more economical and cost-effective than a conventional spark ignition (SI) engine vehicle. Hence, it is necessary to educate customers regarding EV cost-effectiveness and what OEMs are doing. It is also necessary that the Government of India should make people aware of such benefits with mandatory and strict actions.

5. India has completed a different ecosystem and infrastructure altogether and needs a different approach to move toward helping the environment change. Here the major issue is road conditions, which are favorable in some regions and not others in the same city. The limited supply of electricity for such a huge population needs to be managed, and small chargers at residents and stations can easily be installed. So, to save time, precharged batteries can be put in the charging station, which, along with the ease of replacing the battery, will also save time. This is not feasible in four-wheel vehicles, but is possible in two-wheel vehicles. This will work because India is the highest consumer of two-wheel vehicles with 79% market capture (Figure 13.4). Focusing more on the development of two-wheel EVs with the technological benefits of four-wheel vehicles (i.e., regenerative braking, battery swapping technology, hybrid systems, etc.) in an affordable context can revolutionize the Indian EV market. This could change the consumer mindset toward EVs and charging infrastructure, which is already ready for such challenges.

13.2.2 Charging Infrastructure

The Ministry of Heavy Industries and Public Enterprises also sanctioned 241 additional charging stations in September 2020 (Ministry of Heavy Industries and Public

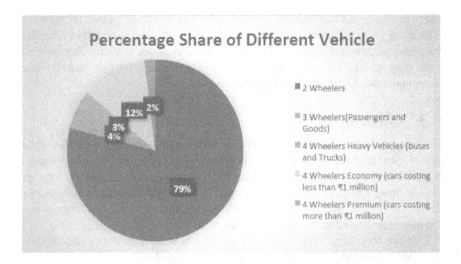

FIGURE 13.4 Latest Indian auto industry market percentage shares. (From NITI-Aayog and World Council, 2018.)

Enterprises, 2019; 2020a,b). DHI also invited proposals for the construction of 2877 EV chargers on highways and expressways in October 2020. Recently, the Ministry of Power announced that it would set up a charging infrastructure across 69,000 petrol pumps in India (Figure 13.5).

Charging infrastructure is a major part of using EV technology. Over the last two years, there have been major developments by OEMs and the Government of India to move toward the installation of fast chargers in major cities in the 12 major states. This helps in bringing new start-ups to both EV and charging infrastructure development. The electricity requirement of charging stations is high with limited supply,

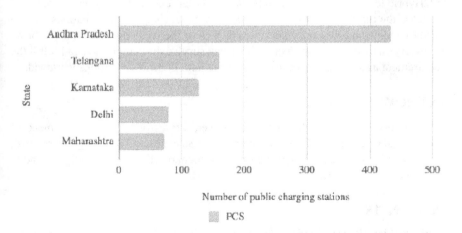

FIGURE 13.5 EV superfast charger installed status by Indian substates. (From NITI-Aayog and World Council, 2018.)

and better solutions can be incorporated. India lies in the subequatorial region and solar energy instead of nonconventional sources can be incorporated in rural highways and mountain terrains to easily power EV charging stations. Similarly, different and new EV charging solutions may enter the market that include battery swapping technology for light-weighted vehicles.

13.3 CONCLUSIONS

EVs have many benefits, but there are limitations, as discussed earlier. New battery technological advancements incorporating hydrogen fuel cell technology, ultra-capacitors, or some other source are at the developing stage. Hydrogen fuel cell cars and their independent charging system are being developed by Hyundai. Eliminating and improving the infrastructure can easily change the ecosystem, and with individual responsibility and understanding of such a market we can create the necessary changes.

EVs are alternative transport utility vehicles both in passenger and heavy vehicles until and unless an alternative source is developed with much more promising results. With a longer life cycle and improved benefits, EVs can be introduced into the market, but the method of implementation should be changed as per the state and condition of that market.

Nations are planning and building their ecosystems to change their transportation needs to EV solutions completely in 5–10 years like Norway (part of Europe) and by 2030 like China. Also, in India NITI Aayog firmly called for a nationwide all-electric three-wheel adoption by 2023 and the two-wheel segment by 2025. These steps help us to understand that nations are ready for the change that is required by environmental conditions.

Technologically, EVs are advanced enough to be able to fulfill the requirement of current generations because we see corporations like Tesla build nationwide ecosystems in the United States, which is a great example of understanding how EV solutions could change transportation and that the idea is feasible and can be successful. Hybrid technology and more integration of high-density battery technology will decrease the limitations of EVs. This idea is a game changer for all nations, but we need to understand the consumer needs as well as the challenges they face to bring necessary changes to the product so that it can survive the competition and fulfill the requirement as an alternate transportation medium in different nations worldwide.

ACKNOWLEDGMENTS

The authors wish to thank Sandeep Grover, Professor of the Department of Mechanical Engineering, J C Bose University of Science and Technology, Faridabad, India, for his guidance, support, and motivation and NITTTR, Chandigarh, India, for providing an opportunity.

REFERENCES

Mariordo (Mario Roberto Durán Ortiz), 2020. Comparison of Plug-in Electric Car Ownership Per Capita in Selected Top Selling Countries and Regional Markets As Plug-in Cars Per 1,000 People.

Ministry of Heavy Industries and Public Enterprises, 2019. Scheme for Faster Adoption and Manufacturing of Electric Vehicles in India Phase II (FAME India Phase II). New Delhi, India.

Ministry of Heavy Industries and Public Enterprises, 2020a. Expression of Interest Inviting Proposals for Availing Incentives under FAME India Scheme Phase II for Deployment of EV Charging Infrastructure on Highways/Expressways, F.No.1(06/2019-NAB-II (Auto) (21349). New Delhi, India.

Ministry of Heavy Industries and Public Enterprises, 2020b. In Phase-II to Fame India Scheme 2636 EV Charging Stations Sanctioned. New Delhi, India.

NITI Aayog and World Council, 2018. Zero Emission Vehicles (ZEVs): Towards A Policy Framework. New Delhi, India.

14 Power Loss Optimization in a Flat Belt Drive

Pratyay Choudhury, Rajesh Kumar,
and Surjeet Singh

CONTENTS

14.1   INTRODUCTION AND RECENT DEVELOPMENTS

Belt drives have many industrial, automotive, and household applications. Flat belt drives are used to transmit power from a driving pulley to a driven pulley by friction between the belt and the pulley. They consist of a tension-carrying cord enveloped by a sheath of rubber or fiber-reinforced rubber. Flat belts are used in drives because of their flexibility, smooth operation, and easy maintenance.

Flat belts have been fairly successful because of a wide range of advantages that they deliver to the performance of the belt-drive model. Flat belts in general make very little noise and absorb more torsional vibration from the system than either V-belt or gear drives. Due to all these reasons, flat belts continue being used in many industrial applications. Even though flat belts are not desirable on small-diameter pulleys, there is no denying that flat belts have many advantages that incentivize their use in industrial applications such as high efficiency (almost as that of gears), high load carrying capacity, less operational noise compared with V-belts, and better shock-absorbing abilities.

Flat belts were prominent during the course of industrialization (18th and 19th centuries) with the establishment and spread of mechanical power systems based on wind, water, and steam engines. In general, flat belts contain tension-carrying cords enveloped

DOI: 10.1201/9781003203681-17

by the upper and lower sheaths, which give the belt a distinct appearance. It is the tension-carrying cords that actually help transmit power from the driver to the driven pulley. In certain theories the surrounding sheaths of belt material have been found to play a crucial role in power transmission, which will be discussed in subsequent sections.

Due to several factors involved in the process of power transmission, a great deal of input power is lost that should have reached the output pulley (follower). The total power loss in a flat belt drive consists of several individual components that can be attributed to belt bending, belt hysteresis, speed losses, torque losses, and aerodynamic drag losses. These components of the total power loss can be described by mathematical equations and can be analyzed with the help of certain analytical tools. The mathematical equations that represent the losses incurred have certain terms that are common to every component of power loss. These common terms include the belt tensions on the tight and slack sides, the belt's material properties, longitudinal stiffness, etc.

The types of belt drives/drive mechanisms are explained using a flowchart shown in Figure 14.1, which, in the subsequent sections, is also discussed in detail:

1. *Flat belts:* Flat rubber belts were developed around the turn of the century primarily as replacements for leather belts. Recent developments in flat belt technology have overcome their previous drawback of high tension. New designs and advances in materials have made both low- and high-power transmission practical and cost efficient, and at speeds that usually exceed other belt designs. Figure 14.2 depicts the flat belt and its geometry.
2. *V-belts:* These belts resemble a trapezoidal lamina when viewed from the sectional plane. The V shape of the belt interior helps it to wedge into the pulley sheaves, which gives it extra durability and operational ruggedness. The V-belt geometry is depicted in Figure 14.3.
3. *Synchronous belts:* These belts are also called timing belts. They were developed to limit the lag between the driving and driven pulleys, which resulted in heavy nonsynchronous movement of the belt drive. The phenomena of slip have been curtailed to a large extent due to the development of these types of belts.

Several analytical models of belt drives are available in the literature. The earliest study of a belt wrapped around a fixed pulley (Beer et al., 1997) and study of frictional mechanics of belt drives under steady operating conditions provided an introductory lesson to belt drives that has been exhaustively covered by prominent

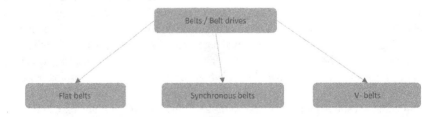

FIGURE 14.1   Types of belt drive commonly in use.

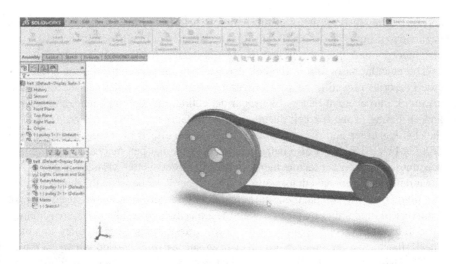

FIGURE 14.2   Schematics of a flat belt drive.

researchers. Two theories based on different belt pulley assumptions were used to model belt drives: creep and shear.

In the creep model, as explained in Beer et al. (1997), belt friction mainly depends on the relative motions of the belt lamina in contact with the pulley surface. The entire contact area is divided into two zones: adhesion and sliding. The sliding zone is responsible for moment transmission through friction forces, while the relative friction between pulley surface and belt material remains zero in the adhesion zone. A comprehensive review of this theory has been done (Fawcett, 1981). The belt tension variations were studied in detail (Della Pietra and Timpone, 2013) using experimental investigation taking both the shear and creep theories into account. Recent inductions of other factors like belt inertia have been studied (Bechtel et al., 2000),

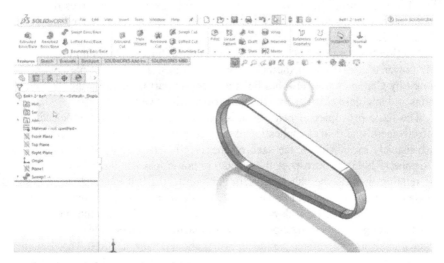

FIGURE 14.3   Schematics of a V-belt drive.

and further analysis taking into consideration the effect of longitudinal stiffness has been carried out (Rubin, 2000). Many new parameters like bending stiffness were also incorporated and duly studied (Kong and Parker, 2005).

The calculation of slip in the belt drive taking into account both the shear and creep theories proved to be very instrumental in explaining flat belt mechanics. It also took into account factors like shear deflection, radial compliance, and seating and unseating of the flat belt during operation (Gerbert, 1991). This not only used the classical creep and shear theories to analyze the effects of seating and unseating of a flat belt drive, but it also helped understand the flat belt mechanics taking into account the many other factors mentioned earlier. Studies to investigate the effect of shear deflection and axial deformation in the rubber layer between the pulley and the belt fibers on the rotational dynamic behavior of the transmission were used to make a numerical model for comparison with the experimental results that were obtained (Tonoli et al., 2006). The study of the power transmission from a pulley to a thin, elastic flat belt in the creep region had been successfully carried out by Belofsky (1973). Using this, the frictional coefficient was calculated, which was found to be inconsistent with the classical Euler's equation. Using the experimental setup, a solution for the slip region was proposed. This proved to be one of the earliest works on belt tensions using an experimental setup providing mathematical solutions.

In the shear model, shear strain in the belt envelope covering the tension-bearing cords is the controlling factor in the mechanics of belt action and performance in the absence of elastic extension (Firbank, 1970). The shear model states that it is the static kinetic forces that actually transmit the bulk of power in the belt transmission as opposed to the kinetic friction forces as conceived in the creep model. This theory is called the shear theory because it states that the friction force in the adhesion zone is determined by the shear deformation, i.e., the relative tangential motion between the pulley surfaces and tension-bearing cords. This shear model mainly caters to the mechanics of a stiff model that can be taken as an inextensible string. The mechanics of a belt lamina sliding over one another to transmit power makes the friction parameters and assumptions governing the elastomers very different, making the shear model different from the creep model.

Friction between the belt surface and the pulley surface plays an integral role in the power transmission in the belt drive mechanics. The contact parameters proposed by Čepon et al. (2010) has been an integral part of assimilating the belt tensions and angular velocities into the numerical belt drive model.

The basic principles of belt transmissions pertaining to energy efficiency, cost-effectiveness, and field of application of respective belt drives including flat belts were exhaustively explained for the sake of energy efficiency studies (De Almeida and Greenberg, 1995). In retrospect, the avenues of power losses in flat belts were explored by relating them to the belt tension, in addition to the belt material properties and angle of contact (Childs and Parker, 1989). In addition, the study of input parameters in any optimization process plays a crucial role in deciding the outcome of the same. In machine design, the optimization process for machine parameters plays a vital role in knowing the output that each parameter brings about (Deihimi, 2009). The elastomeric belt hysteresis losses in a poly V-belt are effectively a function of several factors like dynamic bending, stretching, shear, flank, and radial compression of the belt rubber

(Silva et al., 2018). This led to the determination of hysteresis power losses in the belt drive which, in turn, opened up many new dimensions to belt drive analysis.

Belt drives provide the optimum conditions for power transmission in both industrial and household applications. The relative sliding of the belt on the pulley surface resulting in speed losses also plays an important role as far as power transmission losses are concerned (Balta et al., 2015). In this study, the relation between belt drive parameters and speed losses were obtained by response surface methodology (RSM) and they were found to be in good agreement. The individual effects of belt drive parameters on speed loss are determined using the one-factor-at-a-time (OFAT) test method. Paying attention to the fatigue and wear and tear of machine components is necessary to achieve optimal outcomes. Thus, the investigation of the wear and tear of belt drives to increase the service lifetime of transmission belt drives has to be studied thoroughly (Kumaran et al., 2020). These results have been captured in Table 14.1.

14.2 POWER LOSSES IN BELT DRIVES

Power loss pertaining to a belt drive represents the loss in power transmitted between the input and the output (i.e., the driver and follower pulley). It is the difference between the input power and the output power that takes place due to a host of factors that will be discussed in subsequent sections. Power loss can be attributed to different factors for different belt drive types depending on the geometry and characteristics of the belt drive in consideration. Factors affecting the power loss in a flat belt drive may or may not be considered while analyzing the power loss in a different type of belt drive like V-belts, synchronous belts, and timing belts, which have a different approach to be taken altogether. This mandate a very careful approach to be taken in order to analyze them for the sake of better understanding of these belt drive mechanisms.

The total power loss associated with the belt drive can be divided or segregated into several individual components that depend on several factors regarding the physical conformities to which it has to adhere. The various components of the total power loss are depicted in Figure 14.4.

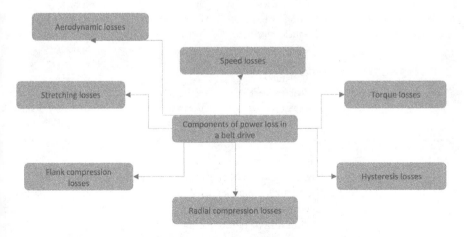

FIGURE 14.4 Different components of power loss in a flat belt drive.

TABLE 14.1

Concise Tabulation of Research Papers with Their Study Objective

Sr No.	Author	Parameters Taken	Focus of Study	Remarks
1	Beer et al. (1997) (Refurbishing Euler's equation)	Belt tensions, coefficient of friction, angle of contact	Ratio of belt tensions along the arc of contact	One of the earliest works on belt drives
2	Fawcett (1981)	Shear deflection of belt envelope, extension of tension cords of the belt	Belt drive mechanics	Review of shear and creep models
3	Della Pietra and Timpone (2013)	Tension in tight and slack sides of the flat belt	Analysis of belt tensions	Comprehensive study of belt tension
4	Bechtel et al. (2000)	Extension of belt fibers, angle of contacts	Investigated the stretching and slipping of belt fibers	Torque transmission of a belt is studied by including the centrifugal acceleration
5	Belofsky (1973)	Belt tensions, belt elasticity, flexural rigidity	Frictional power transmission	Estimation of power transmission of belt drives using belt tensions
6	Childs and Parker (1989)	Belt tensions, longitudinal stiffness, contact angle in slip zone	Power transmission by flat and V-belts	Investigated into power transmission of belt drives with emphasis on extension of tension cords
7	Firbank (1970)	Belt creep, belt tensions in tight and slack sides	Effect of shear strains in determining belt drive	Propagated the shear theory of belt drive mechanics
8	Silva et al. (2018)	Dynamic bending, stretching, shear, flank, and radial compression of the belt rubber	Power losses in V-belts	Modeling of power losses in V-belts using the hysteresis phenomenon
9	Kumaran et al. (2020)	Belt tension, belt speed, power transmitted	Service lifetime of power transmission flat belts	Assessment of flat belt in power transmission

The components of power loss have been explained in detail in Table 14.2 for in-depth understanding of the various concepts involved. The components of power loss explained in Table 14.2 encompass both the flat belts and V-belts and can be used as a source of primary understanding of belt drive mechanics.

TABLE 14.2
Individual Components of Power Losses

S. No.	Component of Power Loss	Basic Definition/Causes
1	Speed losses	These are due to the relative sliding of the belt material over the pulley surface. Speed loss depends on belt creep, flexural rigidity, shear deflections, and loss of belt tensions over one revolution of the pulley.
2	Torque losses	Torque loss, in general, is the difference between the driven and driving torque. It can be attributed to hysteresis loss, friction losses, and idling losses in one complete revolution of the pulley.
3	Hysteresis losses	These are due to the bending and unbending of the belt around the pulleys. They occur at the flexing points, i.e., the points in which the belt comes in contact with the pulley surface and the belt fibers undergo tensile and compressive loading depending on the distance of the belt fibers from the neutral axis of the belt per cycle.
4	Aerodynamic drag losses	They depend on the drag force that is exerted on the belt lamina over one complete revolution. The force on an object that resists its motion through a fluid is called drag. Mathematically, the drag force depends on the velocity and a constant, which depends on the material property of the surrounding fluid.
5	Stretching losses	They occur due to the constrained stretching of the belt polymers/fibers over one complete revolution, which leads to variable tension in the belt as it moves over the pulley surface.
6	Flank compression losses	They occur due to the belt's flank contact with the pulley grooves every time the poly V-belt passes by a pulley. As a result, one side of the flank is unduly pressurized compared with the other side due to the imbalance caused. The net effect depends on the angle of contact and the pulley radius.
7	Radial compression losses	They occur due to the impending pressure caused on the belt fibers, which are cyclically compressed. Every time the poly V-belt passes by a pulley, also due to the pressure generated from the contact between the belt and the pulley, similarly to the flank compression cyclic loading, the belt middle and top layer are cyclically compressed.

The factors affecting the different belt drives have an undeniable connection with the components of power loss enumerated earlier. Thus, it is imperative that the factors affecting the power loss in different belts are discussed holistically. For this purpose, these factors for different belt drives have been tabulated in Table 14.3.

TABLE 14.3
Factors Affecting Different Belt Drives

Important Factors in Flat Belt Drive Mechanics	Important Factors in V-Belt Drive Mechanics
1. Belt tension in tight side	1. Radial compliance
2. Belt tension in slack side	2. Dynamic bending
3. Longitudinal stiffness	3. Bending stiffness
4. Shear deflection	4. Radial and flank compression
5. Angle of contacts	5. Belt tensions
6. Belt fiber shear	6. Flexural rigidity

14.3 POWER LOSSES IN FLAT BELT DRIVE

All the individual components of the total power losses were enumerated earlier. However, not all of them have a role to play in the power loss of a flat belt drive. This is because of the inherent simplicity that is always characterized with flat belt drives, which makes it a compelling study. The power loss components such as flank/radial compression losses, bending losses, and stretching losses are more or less absent in flat belts, which makes the study of flat belts quite simplistic. However, the simplicity in the physical geometry of it is compensated by the overall complicity of it in various industrial and household applications. As was explained in the earlier section, the basic terminologies pertaining to power losses remain the same in flat belts too. The only difference that occurs is in the variance of individual components of power losses.

The components of power losses in a flat belt drive are speed losses, torque losses, hysteresis losses, and aerodynamic drag losses. The rest of the other components of power losses are not very relevant in the flat belt drive model study.

14.4 OPTIMIZATION

Optimization refers to the mathematical process of finding the minimum value of a function. The function to be optimized is called the *objective function*. Usually, most of the optimization problems are minimization problems with the corollary that essentially maximization of a function is actually the minimization of the negative of that function. The variables used to design the function are called *design variables*. Optimization processes are used to obtain the best possible output variable with respect to the boundary conditions provided.

There is no known single optimization method available for solving all optimization problems. Many optimization methods have been developed for solving different types of optimization problems in recent years. The modern optimization methods (sometimes called nontraditional optimization methods) are very powerful and are popular methods for solving complex engineering problems. These methods are particle swarm optimization (PSO) algorithm, neural networks, genetic algorithms, ant colony optimization, artificial immune systems, and fuzzy optimization.

14.5 CONSTRAINED OPTIMIZATION

Constrained optimization problems are problems for which a function $f(x)$ is to be minimized or maximized subject to constraints $\varphi(x)$. A constraint is a hard limit placed on the value of a variable that prevents it from going in certain direction forever. Basically, a constraint places some restrictions on the direction of values that the function variable can possibly take in its domain range.

14.6 UNCONSTRAINED OPTIMIZATION

Unconstrained optimization problems consider the problem of minimizing an objective function that depends on real variables with no restrictions on their values.

14.7 DYNAMIC OPTIMIZATION

Many optimization problems have objective functions that change over time, and such changes in the objective function cause changes in the position of optima. These types of problems are said to be dynamic optimization problems.

14.8 GLOBAL OPTIMIZATION

Global optimization is a branch of applied mathematics and numerical analysis that attempts to find the global minima or maxima of a function or a set of functions on a given set. It is usually described as a minimization problem because the maximization of the real-valued function $f(x)$ is equivalent to the minimization of the inverse of the same function. Global optimization is distinguished from local optimization by its focus on finding the minimum or maximum over the given set, as opposed to finding *local* minima or maxima. Finding the global minimum of a function is particularly very difficult as its conception and application is hard to achieve.

14.9 LOCAL OPTIMIZATION

Local search is usually defined as a heuristic method for solving computationally difficult optimization problems. A local search algorithm starts with a neighborhood local candidate and then moves to a neighbor solution. Local search algorithms move from solution to solution in the space of candidate solutions (also called the search space) by applying local changes, until a solution deemed optimal is found or a time bound is elapsed. Thus, it simply searches the optimum solution in its immediate neighborhood and moves on with the second best candidate in an iterative manner.

14.10 PARTICLE SWARM OPTIMIZATION

The PSO is a novel population-based stochastic search algorithm and an alternative solution to the complex nonlinear optimization problem. It basically tries to optimize a problem by iteratively trying to improve a candidate solution with regard to the

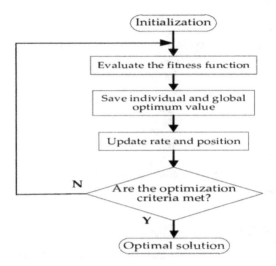

FIGURE 14.5   Flowchart depicting the working procedure of PSO.

given measure of a quantity as depicted in Figure 14.5. It solves a problem by having a population of candidate solutions (or particles) and moving these particles around the search space according to the mathematical formulae given for the particle's position and velocity.

PSO is categorized as a meta-heuristic method as it makes very few assumptions about the function being optimized and can search large population spaces of candidate solutions. Also, PSO does not use the gradient of the function being optimized, which directly means that it does not need the function to be differentiable in the domain range of the variables.

The PSO algorithm was first introduced by Drs. Kennedy and Eberhart in 1995, and its basic idea was originally inspired by simulation of the social behavior of animals such as bird flocking, fish schooling, and so on. It is based on the natural process of group communication to share individual knowledge when a group of birds or insects search for food or migrate in a searching space, although all birds or insects do not know where the best position is. From the nature of the social behavior, if any member can find out a desirable path to go, the rest of the members will follow quickly.

The PSO algorithm basically learned from animal's activity or behavior to solve optimization problems. In PSO, each member of the population is called a particle and the population is called a swarm. Starting with a randomly initialized population and moving in randomly chosen directions, each particle goes through the searching space and remembers the best previous positions of itself and its neighbors. Particles of a swarm communicate good positions to each other as well as dynamically adjust their own position and velocity derived from the best position of all particles. The next step begins when all particles have been moved. Finally, all particles tend to fly toward better and better positions over the searching process until the swarm moves to close to an optimum of the fitness function. This has been lucidly depicted in Figure 14.5, where the input variables have been denoted in the x- and y-axes and the algorithm has searched the swarm population only to find the optimized value.

The PSO method is becoming very popular because of its simplicity of implementation as well as its ability to swiftly converge to a good solution. It does not require any gradient information of the function to be optimized and uses only primitive mathematical operators. Compared with other optimization methods, it is faster, cheaper, and more efficient. In addition, there are few parameters to adjust in PSO, which makes PSO an ideal optimization problem solver in optimization problems. PSO is well suited to solve the non-linear, non-convex, continuous, discrete, integer variable type problems.

The PSO algorithm is a multi-agent parallel search technique that maintains a swarm of particles and each particle represents a potential solution in the swarm. All particles fly through a multidimensional search in the search space in which each particle is swarming around as per its own experience and that of neighbors.

As is evident from Figure 14.6, the PSO algorithm optimizes the value of the input function with successive iterations. The program runtime depends on the number of iterations performed and the swarm size. The objective function is, thus, optimized using PSO. The best fitness value denotes the value of the objective function, which changes with successive iterations as denoted earlier. The numerical values in the axes might change as per the function and boundary conditions imposed on the PSO algorithm.

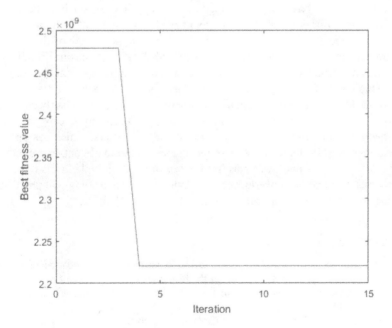

FIGURE 14.6 Variation of best fitness value with iterations performed.

14.11   USE OF PSO FOR OPTIMIZING THE POWER
LOSSES OCCURRING IN A FLAT BELT DRIVE

The mathematical expression that encompasses the entire power loss can be derived from the previously published research papers that will provide a strong base for understanding the fundamentals. The input factors, like belt tension in the tight and slack sides, longitudinal stiffness, modulus of elasticity of the belt material, play an important role in the formation and articulation of the total power loss function (objective function). This equation can be optimized by using the appropriate optimization technique. In this case, the optimization technique chosen is the PSO, as was explained earlier.

The code for this optimization technique can be written in any programming language/code that is coherent with the demands of the problem to be solved.

PSO often produces better results compared with other optimization techniques. This can be attributed to the meta-heuristic tendency of PSO, which nullifies the requirement of a gradient of the function to be optimized. This gives PSO a distinctive edge over other algorithms. Moreover, the simplicity and faster processing time of PSO adds to its credibility. All these factors make PSO an ideal algorithm to be used for optimizing a range of functions, which no other optimization algorithm can offer.

14.12   CONCLUSIONS

Analytical models based on the geometry of flat belts have shown how the belt fibers behave in the adhesion and slip zones and have been compared with the findings of numerical models, which provide a much more analytical view of the same. Belt parameters such as belt tensions in the tight and slack sides, flexural rigidity, shear deflections, contact angles in both the slip and adhesion zones, and so forth, have been studied to effectively understand the behavior of flat belts. These belt parameters have been fairly represented in the objective function, which has been, in effect, used to optimize the power loss function using PSO. The hysteresis phenomenon has been included in the belt analysis in the past few years, which has opened up new dimensions in the study of drive mechanics. The role of contact arc under slip and adhesion zones and hysteresis has thrown light on multiple facets of belt drive mechanics that had been overlooked for a long time. Such a study will prove helpful during research work to optimize power loss in a flat belt drive.

REFERENCES

Balta B., Sonmez F.O., and Cengiz A., 2015. Experimental identification of the torque losses in V-ribbed belt drives using the response surface method. *Proceedings of the Institution of Mechanical Engineers, Part D: Journal of Automobile Engineering* 229(8), 1070–1082.

Bechtel S.E., Vohra S., Jacob K.I., and Carlson C.D., 2000. The stretching and slipping of belts and fibers on pulleys, *Journal of Applied Mechanics* 67(1), 197–206.

Beer F.P., Johnston E.R., Eisenberg E.R., and Sarubbi R.G., 1997. *Vector Mechanics for Engineers*. McGraw-Hill, New York.

Belofsky H., 1973. On the theory of power transmission by a flat, elastic belt, *Wear* 25(1), 73–84.

Čepon G., Manin L., and Boltežar M., 2010. Experimental identification of the contact parameters between a V-ribbed belt and a pulley, *Mechanism and Machine Theory* 45(10), 1424–1433.

Childs T.H.C., and Parker I.K., 1989. Power transmission by flat, V and timing belts. In: Dowson D., Taylor C.M., Godet M., and Berthe D., eds. *Tribological design of Machine Elements*. Elsevier, Amsterdam, Vol. 14, pp. 133–142.

De Almeida A., and Greenberg S., 1995. Technology assessment: energy-efficient belt transmissions, *Energy and Buildings* 22(3), 245–253.

Deihimi A., 2009. Design optimization of switched reluctance machines for maximum torque/current using BEM-based sensitivity analysis, *International Journal of Recent Trends in Engineering* 2(5), 167–172.

Della Pietra L., and Timpone F., 2013. Tension in a flat belt transmission: experimental investigation, *Mechanism and Machine Theory* 70, 129–156.

Fawcett J.N., 1981. Chain and belt drives—a review, *Shock & Vibration Digest* 13(5), 5–12.

Firbank T.C., 1970. Mechanics of the belt drive. *International Journal of Mechanical Sciences*, 12(12), 1053–1063.

Gerbert G.G., 1991. On flat belt slip. In: Dowson D., Taylor C.M., and Godet M., eds. *Vehicle Tribology*. Elsevier, Amsterdam, Vol. 18, pp. 333–339.

Kong L., and Parker R.G., 2005. Steady mechanics of belt-pulley systems, *Journal of Applied Mechanics* 72(1), 25–34.

Kumaran V.U., Zogg M., Weiss L.,and Wegener K., 2020. Design stand for assessment of flat design of a test stand for lifetime assessment flat belts in power transmission, *Procedia CIRP* 91, 356–361.

Rubin M.B., 2000. An Exact Solution for Steady Motion of an Extensible Belt in Multipulley Belt Drive Systems. ASME. *Journal of Mechanical Des.*, 122(3), 311–316. https://doi.org/10.1115/1.1288404

Silva C.A., Manin L., Rinaldi R.G., Remond D., Besnier E., and Andrianoely M.A., 2018. Modeling of power losses in poly-V belt transmissions: hysteresis phenomena (enhanced analysis). *Mechanism and Machine Theory* 121, 373–397.

Tonoli A., Amati N., and Zenerino E., 2006. Dynamic modeling of belt drive systems: effects of the shear deformations. *Journal of Vibration and Acoustics* 128(5), 555–567.

15 Comparative Analysis of OFET for Low-Power Flexible Electronic Devices Manufacturing

Bodade Sandip Vasudeo, Banoth Krishna,
Parveen Kumar, Sandip Singh Gill,
and Balwinder Raj

CONTENTS

15.1 INTRODUCTION

Organic field-effect transistors (OFETs) use organic semiconductors in their channel instead of a-Si conventional semiconductor material. Figure 15.1 shows that the cost of silicon is higher compared with the cost of organic materials, but there are some different parameters that show that silicon performs better than organic materials. The advantages of organic materials are large area, low cost, low-end flexibility, and easier manufacturing; most researchers and academicians work with organic material. OFET construction is the same as the construction of thin film transistors. There are three preparation methods:

1. Small molecules by vacuum evaporation,
2. Polymer casting or small molecules by solution casting, and
3. Mechanical transfer on a substrate of a peeled single crystalline layer.

Dip coating, spin coating, inkjet printing, and screen printing are included in the solution coating process. The electrostatic lamination method is based on manually peeling off of a thin layer on a single organic crystal. The thickness of the active

DOI: 10.1201/9781003203681-18

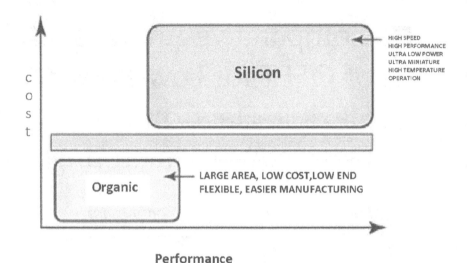

Performance

FIGURE 15.1 Cost versus performance.

layer and the oxide layer of the gate should be less than 1 μm. Silicon dioxide acts as a dielectric layer in gate insulators. Thermally oxidized silicon is used as a substrate for OFETs. The thermal evaporation and coating from the organic solution of these two techniques create a polycrystalline active layer; it is easy to generate but gives poor transistor performance. The OEFT functions as a capacitor. One capacitor plate as a conducting channel between two ohmic contacts, which are the main two OFET terminals (called source and drain), and another capacitor plate to regulate the induced charge into the channel called gate, which is the third OFET terminal, is shown in Figure 15.2.

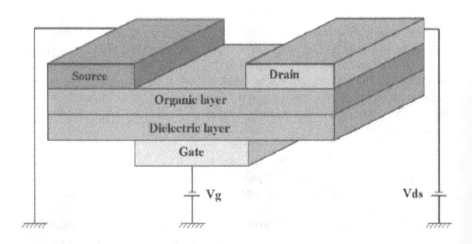

FIGURE 15.2 Structure of field-effect transistor.

15.2　THERE ARE TWO TYPES OF OFET STRUCTURES

The studies reported on OFET structures are classified into main two types (Figure 15.3):

1. Top contact
2. Bottom contact

The OFET gets its design architecture from an inorganic metal oxide semiconductor field-effect transistor (MOSFET). It has three main components, source, drain, and gate, along with a dielectric material layer and the active semiconductor layer. The MOSFET device has two basic designs, as shown in Figure 15.3. The device configurations are top contact and bottom contact. These two top contact devices perform better because of reduced contact resistance and increase in the charge injection area. Figure 15.3 shows that in the top contact configuration source and drain terminals are placed over the organic layer, dielectric layer, gate, and substrate in chronological order. In the bottom contact configuration source and drain are placed in the organic layer, dielectric layer, gate, and substrate.

Chong Di et al. (2012) proposed that the sensor networks, which are used to detect efficient various stimuli that are related to biological, Internet of Things (IoT), wearable electronics, and more specific environmental applications, use various flexible and stretchable electronics devices. Sensors are used to measure pH, glucose, temperature, pressure etc. Wei Shi (2020) proposed that the IoT and

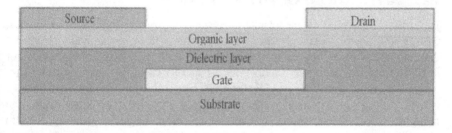

Top Contact

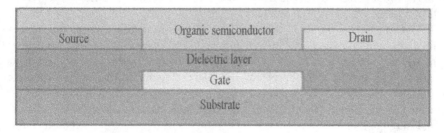

Bottom contact

FIGURE 15.3　Top contact and bottom contact.

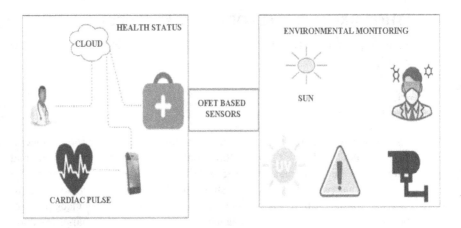

FIGURE 15.4 Broad application area of flexible electronics.

wearable flexible devices are a part of our day-to-day life, and recent technological advancements are done in the IoT (Figure 15.4). Wearable electronics will achieve maximum market share up to 2024. Due to the solution processibility and flexibility as well as chemical properties of organic optoelectronics, materials show different properties like light emission, charge transport, photovoltaic, thermoelectricity, and superconductivity.

15.3   APPLICATIONS OF FLEXIBLE DEVICES

By virtue of organic optoelectronics material, the organic light-emitting diodes, OFETs, and organic solar cells are designed in these three multifunctional organic devices, which are mostly used by industries, as well as in consumer products, or flexible electronic devices such as electronic paper RFID cards, biosensors, and after long research in organic devices. We have achieved fabrication of high-performance organic devices with high-effect mobility along with good association between molecular structure and device performance by using one device with one or more physical properties. Such devices can be extensively used by researchers in the development of OFET-based sensors, photodetectors, memory cells, optical transistors, and biomimetic electronic skin. Body temperature is the main parameter for diagnosing health issues.

Wei Shi (2020) proposed that glucose sensors are helpful in detecting glucose levels in humans. The prices of actuators and sensors have fallen because they are mostly used in the field of IoT in the Internet-related activities. Sensors and actuators are also used in wearable electronics, and the compound annual growth report (CAGR) says that the use of sensors and actuators are increasing in day-to-day life, achieving 6% annual growth by the year 2020. Semi-autonomous cars, IoT products, and intelligent embedded controls are the future of the electronics industry. Yoon Hoo Lee et al. (2017) proposed that OFETs are used in sensing applications and the active semiconducting layer is exposed to the external environment to detect or sense different parameters of the sensors. Chemical sensors are used to monitor

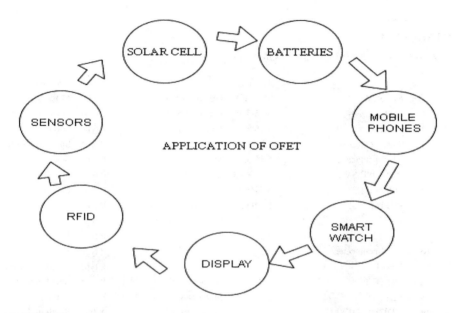

FIGURE 15.5   Different applications of OFETs.

environmental changes such as pollution, toxic contents in air, and water pollution. Yaping Zang (2015) proposed that a magnetic property of inorganic semiconducting materials such as giant magnetoresistance and the Hall effect of OFETs are used as ultrasensitive magnetic sensors. P. Cosseddu et al. (2012) proposed that flexible devices should be used to analyze the parameters of joint movement and breath rate for biomedical systems. The functionality of these flexible devices depends on the cascaded structure shown in Figure 15.5.

15.4   RESEARCH BACKGROUND

Electronic products used under the category of wearable and flexible have greatly evolved in the last decades. These devices are used throughout the world and there is more research being done to develop these devices as portable and wearable devices, such as wristwatches, smartwatches, and pocket watches (Yuvaraja et al., 2020) The current advancement in this technology is toward the blending of devices and OFET-based flexible sensors, including chemical sensors, light sensors, biosensors, and pressure sensors. The material with HOMO and LUMO gaps is given in Table 15.1.

15.5   RESEARCH DISCUSSION

Following numerous innovations in OFETs over the past decade, the system features of OFETs are now equivalent to those of a-Si FETs. These advances have been made, and now OFETs are allowed to be used in true, low-cost electronic applications, and differentiated devices, such as single device applications, have been

TABLE 15.1
Materials with HOMO and LUMO Gap

Sr No.	Material	HOMO/ LUMO	Deposition Method	Type	Polymer/ Molecule
1	Pentacene	5.16/3.35	Evaporation	P type	Small molecule
2	P3HT	5.16/3.2	Spin coating	P type	Polymer
3	TIPS-pentacene	5.2/3.14	Drop coating	P type	Small molecule
4	TIPS-pentacene	5.1/3.4	Spray coating	P type	Small molecule
5	P-29-DPP-SVS	5.35/3.98	Spin coating	P type	Polymer
6	DPP@T-TT	5.31/3.3	Bar coating	P type	Polymer
7	PDFDT	5.24/3.37	Bar coating spin coating	P type	Polymer
8	Ph5t2	5.85/–	Mechanical probe	P type	Singing crystal
9	PQT12	5.24/2.97	Electro spinning	P type	Polymer
10	TIPS-pentacene	5.2/3.14	Drop coating	P type	Small molecule
11	DNTT	5.4/2.5	Evaporation	P type	Small molecule
12	DDETTF	5.36/2.53	Evaporation	P type	Small molecule
13	PDPP3T	5.35/3.81	Spin coating	P type	Polymer
14	NDI3HU-DTYM2	6.38/4.38	Spin coating	N type	Small molecule
15	NDI(2OD)(4tBuPh)-DTYM2	–	Spin coating	N type	Small molecule
16	Dif-TESADT	–/–	Picofilter fluidic dispenser	P type	Small molecule
17	PIDT-BT	5.23/–	Spin coating	P type	Polymer
18	PTAA	5.25/2.3	Spin coating	P type	Polymer
19	PTA	5.33/2.04	Evaporation	P type	Small molecule
20	CuPe	5.2/3.5	Evaporation	P type	Small molecule

increasingly explored for the last couple of years. Thus, the study of multifunctional OFETs, which includes multifunctional material design, condensed structure modulation, application geometry optimization, functional modification interfaces, and the integration of devices, should have continued attention in the future and open new markets for apps and applications. Speedy growth in this field will make a huge contribution to a deeper understanding of the link between molecular structure and property. Thus, from the previous observations, in Table 15.1, we have observed the HOMO and LUMO energy gaps for different materials used in sensors

15.6 CONCLUSIONS

This study presents the OFET and its advancement in the design of flexible electronics. The concentration of this work is focused on the evolution of parameters reported in previous research papers. The comparative statement is developed based on different materials used in OFETs. Parameters such as material HOMO, LUMO, type of semiconductors for charge carrier mobility, and current and voltage analysis have been developed. The various applications of OFETs in sensors and flexible electronics devices are also discussed.

ACKNOWLEDGMENTS

The authors wish to thank the Director of DTE Maharashtra and Principal, Govt Polytechnic, Jintur. Dist. Parbhani Maharashtra, Head of The Department Electronics and Communication Engineering, and all faculty members of Electronics and Communication Engineering Department and supporting staff of NIITTTR Chandigarh, India.

REFERENCES

Cosseddu P., Loi A., Basiricò L., Lai S., and Bonfiglio A., 2012. Organic bendable and stretchable field effect devices for sensing applications. SENSORS, 2012 IEEE, Taipei Taiwan. doi:10.1109/ICSENS.2012.6411281.
Di C.A., Zhang F., and Zhu D., 2013. Multi-functional integration of organic field-effect transistors (OFETs): advances and perspectives, *Advanced Materials* 25(3), 313–330. doi: 10.1002/adma.201201502.
Lee Y.H., Moo M.J., Lee Y., Kweon O.Y., and Oh J.H., 2017. Flexible field-effect transistor-type sensors based on conjugated molecules, *Chem* 3(5), 724–763.
Shi W., Guo Y., and Liu Y., 2020. When flexible organic field-effect transistors meet biomimetics: a prospective view of the internet of things, *Advanced Materials* 32(15), 1901493. doi: 10.1002/adma.201901493.
Yuvaraja S., Nawaz A., Liu Q., Dubal D., Surya S.G., Salama K.N., and Sonar P., 2020. Organic field-effect transistor-based flexible sensors, *Chemical Society Reviews* 49, 3423–3460. doi: 10.1039/c9cs00811.
Zang Y., Zhang F., Huang D., Di C.A., and Zhu D. 2015. Sensitive flexible magnetic sensors using organic transistors with magnetic-functionalized suspended gate electrodes, *Advanced Materials* 27(48), 7979–7985. doi:10.1002/adma.201503542.

16 A 360° Industry Process Optimization Using Industry 4.0 Technologies

Dinesh Chander Verma, Nisha Chugh,
Poonam Verma, and Amit Kumar Dubey

CONTENTS

16.1 INTRODUCTION

The term Industry 4.0 refers to the fourth industrial revolution, which is defined as information and communication technology (ICT)-enabled tools and techniques that heavily focus on automation, intelligent interconnectivity, and, real-time data analysis of industrial processes. More precisely, under this recent trend, technologies like the Internet of Things (IoT), machine learning, big data analytics, robotics, augmented reality, virtual reality (VR), and cyber-physical space, and three-dimensional (3D) printing plays a vital role (Bigliardi et al., 2020).

Industry 4.0 was first coined by Germany to introduce the radical shift toward digitization in industrial production and to develop competitiveness among manufacturing industries. Afterward, many developed and emerging countries like the United States, China, Brazil, and India started promoting Industry 4.0 technologies for the day-to-day

DOI: 10.1201/9781003203681-19

processes of their industries (Dalenogare et al., 2018). The main motive of Industry 4.0 is to improve the overall industrial processes such as manufacturing, sales and purchase, human resource management, research and development, supply chain, and so forth, in an intelligently and optimized manner (Vaidya et al., 2018). Further, Industry 4.0-based industries have an edge over the traditional industries in terms of efficiency, productivity, reduced cost, decision making, and overall reputation. These recent practices enable the various processes to communicate and monitor each other on a real-time basis to achieve a high degree of optimization (El Hamdi et al., 2019).

The future of industrial processes has become more autonomous and intelligent with the use of digital systems. Further, there is the concept of knowledge-based industry and self-driven types of factories, which will greatly improve decision making, efficiency, and competitiveness in the industry. It is assumed that with the use of Industry 4.0 concepts the overall productivity of the industry will increase by 35%–40% (Zhou et al., 2016). Industry 4.0 is the need of the hour to enable the machines to work in self-learning, self-monitoring, self-control, and self-aware modes.

In this chapter, the various Industry 4.0 concepts are discussed to optimize the overall processes. The chapter is divided into seven sections: introduction, literature survey, major challenges, Industry 4.0-enabled processes, conclusion, and finally future scope.

16.2 MAJOR CHALLENGES IN THE TRADITIONAL INDUSTRIAL PROCESSES

Before discussing challenges in the industry, there are seven principles for the development including implementation, and maintenance of the existing quality management system. These principles are described by ISO 9000. In the first step, the present and future demand of customers must be identified as they are the most important part of an organization. Second, a manager along with his team should work to achieve the goal of the company. Third, to attain the quality objective, responsibilities must be assigned to each person involved. Along with responsibilities, resources should be determined and provided to attain the goal. Further, data and information should be properly arranged so that quality decisions can be made to improve the performance of the organization and to maintain the relationship between the organization and external stakeholders (Foidl and Felderer, 2016). Figure 16.1 depicts the major challenges faced by the industry.

FIGURE 16.1   Challenges faced in industry in different sectors.

16.2.1   MANUFACTURING INDUSTRY

The manufacturing industry is one of the major revenue-producing segments of universal wealth and plays an important role in producing materials needed by various companies across the globe. Figure 16.2 represents the major area in which Industry 4.0 integration can enhance the manufacturing process.

1. *Scarcity of expert personnel:* In the traditional age, most of the work was done manually, but in the digitized age, most tasks have become automated. Thus companies require employees with special skill sets. As the automation of tasks and machine sensors has become commonplace in industry, a large quantity of data need to be created by manufacturers. Advanced skills and innovative ways of thinking are needed so that complicated work can be performed faster. Existing workers have to find their feet in the updated working environment. This is a cause of resistance for current employees to leave their existing operational traditions (Trstenjak, 2018).
2. *Project management:* It is a process in which different departments of an organization concur together to achieve objectives and solutions to a problem (Bag et al., 2021). In addition to this, the projects in the industry are not flexible in terms of time, economy, and quality, thus projects need to be monitored firmly. Such rigidity means less adaptability to make any changes during project growth. Thus, it becomes quite difficult for the team who wants to construct the best product but is bounded by deadline constraints.
3. *Data analysis and integration:* In today's data-driven world, there are diverse ways to produce data. It can be from machine sensors, process data, quality data, communication data, and many more; all these are combined to form an outburst in data volume. With the increase in the volume of data, new methodologies and technologies are required to store, process, and manage the data. Ground-breaking algorithms and techniques are required to use and expand gain from data. To analyze data and to maintain relationships between the information available, there is a high demand for data

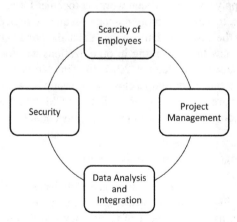

FIGURE 16.2   Challenges faced by manufacturing industries.

engineers by industries. In addition, no standard approach is available to manage data. Similar data are available with sectors of companies in various formats leading to inconsistency and hence inefficient decisions by the company, thus, decreasing performance (Abraham et al. 2016).

4. *Security:* This is a major concern today and would also be in the future for industries. All companies require their employees, manufactured goods, manufacturing facilities, and data to remain secure. With the tremendous increase in usage of smartphones, tablets, and industrial machines, keeping an eye on these devices has become a major challenge from a hardware and software perspective. On the one hand, it is making our day-to-day life easier, and on the other hand, it is increasing security risk, resulting in the possibility of manufactured products having viruses. This may also result in a heavy penalty for company returns (Abraham et al. 2016).

16.2.2 SUPPLY CHAIN

The term supply chain management (SCM) defines the sequence of steps involved to deliver the product to the customer. It involves various stages like production, development, marketing, and transportation such that the final product can be received by the consumer when ordered. SCM makes development series well-organized and at a lower expenditure. Figure 16.3 represents the subprocess involved in the supply chain process.

FIGURE 16.3   Sequence of steps involved in the supply chain.

There are two stones on which SCM is based. First, the delivery of products to customers is a united effort of numerous organizations involved. Second, as the organizations in the supply chain have been working together for ages, there is a need to think out of the box to achieve a more managed sequence of activities to deliver the finished product to the customer. During the 1980s, the bond between consumer and supplier took on a new face, resulting in organizations developing strong relationships with clients (Awad and Nassar, 2010a,b). Thus, to achieve a more managed sequence of activities and to develop a close relationship between strategic and logistic customers, there are some challenges faced by SCM, which are discussed below. In Figure 16.4, major challenges faced by the supply chain have been depicted.

16.2.2.1   Challenges Faced in Supply Chain

1. *Integration:* Supply chain integration generally includes all steps of the supply chain starting from receiving a customer request to delivering the finished product. With the advancement of technology and digitization of business, the supply chain has become a matter of concern of high-level managers (Awad and Nassar, 2010a,b). Integration of the supply chain helps to reduce

FIGURE 16.4 Major challenges in the supply chain.

the snag caused by various factors like transaction cost, planning and creating strategies for business, and having operational principles and flexibilities. Integrating customers and organizations and creating a strong bond between them, and keeping account of all technical, order requests, production, manufacturing, marketing, and sales aspects can bring an accountable increase in sales (Abdirad and Krishnan, 2021). In addition to the challenges previously mentioned, integration challenges can be broadly classified into three sectors: micro-environmental challenges, which include challenges concerning strategic planning management, operation flexibility, and procurement management; business macro-environment, which includes challenges regarding supplier competence and effect of globalization; and technical challenges like an integration of data, information, and application. In Figure 16.5, the major aspects that need to be adhered to have been shown.

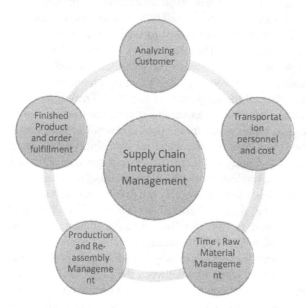

FIGURE 16.5 Major aspects of the supply chain.

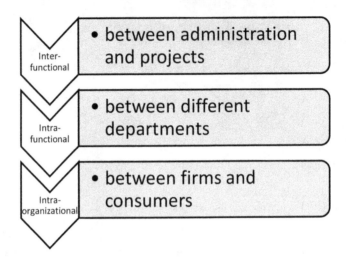

FIGURE 16.6 Components of effective operation.

2. *Operation:* Scaffold of SCM relies on three aspects: synchronization between administrative activities and processes of organization (i.e., intra-functional); synchronization between different sectors of the organization that are interfunctional; and synchronization between different supply chain activities that are held between organizations, their suppliers, and consumers (Ballou et al., 2000) as shown in Figure 16.6.

   To maintain coordination between interfunctional, intrafunctional, and organization, there needs to be a close relationship between customer and firm (Salam, 2017). If the demand for a product is increased overnight, then high-level managers need to maintain a balance between demand and supply. Additionally, it is necessary to take into account consumer conduct (Kumar et al., 2020). The challenge of operational management can be attained by achieving mutuality between trust and technology.

3. *Purchasing:* While working with the supply chain, there is a great need to bring together all the steps that are involved. There is no defined standard of synchronization between various events, but short of synchronization it may lead to an imprecise prediction of inventory; poor customer service; improper stock expenditure, and time availability of stock in the market that results in a delay in order completion request, resulting a less satisfactory product (Arshinder et al., 2008). Purchasing is a crucial step in the supply chain, it has a long history. By the 2000s, an Institute of Supply Chain, earlier known as the U.S. National Association of Purchasing Management, was established for the dazzling inclusive impact of purchasing (Chen and Paulraj, 2004). Also, the amendments in the area of purchasing from traditional and least efficient ways to stratagem and external effectualness require the deliberation between the organization and its network in the value chain were quite challenging. Thus, extending the field of purchasing was not only meant to provide an efficient solution or minimum

Receiving raw material and matching invoice

Finding demands

Purchasing

Putting in a request for purchase

Placing order

Getting approved

FIGURE 16.7 Critical aspects of the purchase.

cost product but also meeting business demands and satisfying customer demand (Gundlach et al., 2006). Moving from labor to rising technologies and the traditional way of purchasing and supply chain influenced the role of purchasing. In Figure 16.7, major aspects of purchasing are presented.

4. *Distribution:* The next step of the supply chain is distribution. While considering the distribution phase, the objective of the firm is to deploy the goods on time in high-demand areas, hoist remuneration, or hire temporary workers so that finished products can be delivered fast and to train personnel to work efficiently with e-commerce. Shifting from the traditional way of distribution to a virtual mode through e-commerce was a challenge for both employees and customers (Zhao et al., 2005). According to research, in India, almost 40% of the grain is wasted due to less efficient management of the food supply. As distribution is the time taken to process, ensuring smoother delivery of the product at a nominal price is a major concern (Mangla et al., 2019). The distribution channel in India is customary and inimitable because it consists of a retail, wholesale, and logistics infrastructure. Managing channels in every situation, especially when the market is down, is quite challenging. Thus, skilled personnel is required who can work on promotion of the products through e-commerce and manage all that goes with it (Mulky, 2013). Training employees to work with emerging technology and also training customers to purchase the product using these technologies through which they place an order, make a payment, and track a shipment to find the delivery time was a challenging situation. As shown in Figure 16.8, the bottom line is there can be difficulties in building digital skills and keeping oneself updated, fulfilling user needs, training customers, managing technical resources, and satisfying infrastructure requirements.

FIGURE 16.8   Challenges in distribution.

16.2.3 Research and Development

Industry 4.0 is also known as a smart factory, which refers to the integration of emerging concepts in the manufacturing industry. Also, from an education point of view, research scholars have kept a keen interest in Industry 4.0 to study the impact of automation on traditional processes (Erro-Garcés, 2019). With the advancement in technology, Industry 4.0 is a comparatively new approach in the area of research, and there are many challenges faced in different sectors of industry like those in manufacturing, production, or integration; thus, different case research strategies need to be used (Khan and Turowski, 2016).

16.3   INDUSTRIAL REVOLUTION: BRIEF BACKGROUND

Most Industrial 4.0 companies have started investing in these new transformations, as digital technologies have changed the way of managing sales and purchases, human resource management, research and development, and supply chains to sustain in the competition. In the back office, all of these processes are controlled and managed by a team of highly skilled professionals. It is also assumed that those who do not accept these transformations will no longer be part of the race.

The fourth industrial revolution highly focuses on automation, intelligent interconnectivity, and cyber-physical space, which enables the industry to analyze, monitor, and automate business processes intelligently. Technologies such as big data, IoT, cloud services, and intelligent predictive decision support systems are used to optimize productivity and efficiency (Preuveneers and Ilie-Zudor, 2017). Many researchers have worked toward the optimization of the industrial process with the help of Industry 4.0 technologies. This different generation of industrial revolutions has added some tools and technologies to the previous one. Figure 16.9 depicts these generations along with their integrated technologies.

The first industrial revolution came into existence in the late 1800s, and water and steam power were the driving forces. The second industrial revolution appeared in the late 1900s and the start of the 20th century. Electricity played a vital role in mass

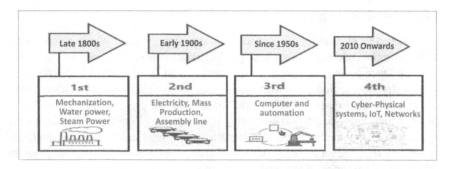

FIGURE 16.9   Industrial revolutions.

production in factories. Electric engines, railroads, and steel industries flourished like never before and it was considered the biggest jump in industrial revolutions (Hwang, 2016).

16.4   OPTIMIZATION USING INDUSTRY 4.0

Most of the industrial processes can be integrated with the Industry 4.0 concepts to achieve a high degree of optimization. In the following section, a few important technology-enabled applications of Industry 4.0 will be discussed for better insight (Pilloni, 2018).

16.4.1   QUALITY CONTROL

Big data is the one major technology that can bring a significant improvement in the quality control process. Advanced analysis techniques are in use more widely to compare and establish relationships between related information and data collected from sensors so that more optimized solutions can be achieved. Many visualization tools are used to compare the quality achieved during the last few years to have better control over the process. There are statistical methods such as advanced process control (APC), statistical process control (SPC), and Six Sigma strategies that can be applied to enhance process control (Sousa et al., 2017).

16.4.2   SUPPLY CHAIN

SCM in the industry was a very challenging process using traditional tools. Industry 4.0-enabled technologies have improved SCM, and tracking has become very convenient. The relevant data can be fetched from the mobile device, IoT, and radio frequency identification technology (RFID) to make the supply chain and logistics process intelligent and smart. RFID tagging is being used to generate a huge amount of strategic and operational data in terms of volume, variety, velocity, value, and veracity for better data mining and future decisions. There are many enterprise resource planning (ERP) systems with multiple modules (e.g. logistic, finance, sale purchase, payroll, and manufacturing) to optimize the industry-specific process (Oghazi et al., 2018).

16.4.3 Manufacturing Control

Manufacturing is directly connected with the supply chain, which refers to the activities associated with the actual product development. The various corrective measures such as defect tracking, testing, and validation of a product are controlled by Industry 4.0-enabled tools. Many industries started adopting the intelligent manufacturing system (IMS) to reform the manufacturing process. Cloud-based, IoT-enabled, and sensor-based manufacturing plays a game-changing role in the overall optimization of manufacturing control. Real-time data are fetched and analyzed to validate the product as per given specifications, even minor diversions from the actual specifications are addressed intelligently.

16.4.4 Design Control

The high-end designs of products are supported by augmented reality (AR) and VR for additive manufacturing (Zheng et al., 2018). Most automobile industries are now highly dependent on autonomous and intelligent robots for quick and high-end products. Computer-aided design (CAD) and computer-aided manufacturing (CAM) integrated with 3D printers and cyber-physical systems (CPSs) is gaining popularity as designing tools to facilitate high-quality prototypes. A good design mechanism always results in a quality product.

16.4.5 Personnel Management

A skilled human resource is considered the most valuable asset for an industry. Keeping this resource up to date in terms of emerging tools and technologies is the key responsibility of an organization. Organizing training, workshops, and seminars are essential activities for the overall development of the industry. In literature, many authors have proposed their ideas for upgrading and accessing the skill set of the human resource. Competency mapping tools are used to find the right person for the right job. Blockchain- and fuzzy logic-based applications are used in industry to recruit and manage the human resource (Onik et al., 2018).

16.4.6 Research and Development

The Industry 4.0 integrated industry cannot survive unless the aptitude of research and development was imbibed in the work culture. Technologies are changing very fast, concepts like IoT, big data, cloud computing, and blockchain applications will be driving forces in the future. The fourth revolution is more about self-driven machines, virtualization, interoperability, real-time analysis, service-oriented applications, and reconfigurability (Cigánek et al., 2018). To keep pace with these emerging technologies and to gain the maximum out of it research-oriented man force is highly recommended. Many tools and simulator are available to carry out research work to optimize the industrial processes.

16.5 PROS AND CONS OF INDUSTRY 4.0

Every coin is two-faced, so implementing Industry 4.0 also has pros as well as cons. Due to the certain downside of technologies, we cannot ignore the expected optimization in overall industrial processes. Guo-jian Cheng et al. focused on the importance of these emerging technologies and emphasized interconnectivity, data integration, and innovations in the industry for maximum gain. A smart or digital factory proposes dynamic production instead of a fixed one, which overcomes the issues related to fixed line (Cheng et al., 2017).

Despite the many advantages of these emerging technologies, only 14% of corporations are ready to adopt the changes by Industry 4.0. It is revealed from the literature that only 4 of 10 companies have shown good progress in implementing these changes, which again varies from country to country. The majority of companies have either no or negligible progress toward implementing these technologies due to a lack of adequate facilities. These barriers in adoption in Industry 4.0 greatly vary from country to country. Developed countries have generally well-defined national policies for industrial development, and the adoption rate is higher; whereas in developing countries the adoption of Industry 4.0 technologies depends on the initiatives of an individual industry rather than national policies(Raj et al., 2020).

Michael Sony discussed the various aspects of Industry 4.0 along with its pros and cons. Implementing Industry 4.0 means integrating IT as an integral part of the product life cycle. There is no doubt that it will change the overall structure of the business for the better in this competitive era, but every technology integration is associated with some side effects. Table 16.1 describes a number of pros and cons in implementing Industry 4.0 (Sony, 2020).

From Table 16.1, it is clear that the implementation of Industry 4.0 is associated with pros as well as some cons. In nutshell, over some time the overhead associated with Industry 4.0 could be suppressed by its benefits. The integration of these systems is reflected in improved monitoring and controlling of the equipment,

---

TABLE 16.1
Pros and Cons of Industrial Revolution 4.0

Pros	Cons
Higher productivity	Highly skilled employees
Increase in operational efficiency	Data security
Better products and services	High setup cost
Competitive advantage	Risk of system failure
Positive customer feedback	Industry and market disruption
Improved strategic prediction	Ethical issues
Better decision making	Partial implementation of Industry 4.0 may lead to failure
Improved profit gain	
Reduced wastage of resources	

---

conveyors, and products. It is also reflected in iterative cycles of feedback that process the big quantity of data (big data) and help in the revision of virtual models with the proper input from the physical processes, resulting in an intelligent factory. Industry 4.0 strengthens the industries from vertical integration, horizontal integration, and end-to-end engineering. Vertical integration enables the strong integration between the production and management levels in the industry. Horizontal integration refers to the strong collaboration between enterprises for real-time information sharing. End-to-end engineering refers to the integration of technologies in the whole product life cycle, i.e., from production until after sales (Dalenogare et al., 2018).

16.6 CONCLUSION AND FUTURE SCOPE

It has been observed that the industry is facing many challenges in its routing processes. Although in many industries processes are being managed traditionally, the desired level of optimization is hard to achieve traditionally. In the last decade many companies in developed countries have adopted the concepts of Industry 4.0 and observed 30%–40% growth in productivity. Now even the progressive countries have shown a keen interest in implementing Industry 4.0 concepts. Despite the few initial cons of Industry 4.0, its integration offers a tremendous perspective toward the sustainability and viability of the systems. The integration of Industry 4.0 means to integrate the technologies like IoT, big data, blockchain, and cyber-physical space into all critical processes to achieve self-driven machines, virtualization, interoperability, real-time analysis, service-oriented processes, and reconfigurability. In a nutshell, future industries can be imagined without the integration of these emerging technologies.

Industry 4.0 is a multidisciplinary field in which experts from different domains are required to solve complex problems. Supply chain planning is a critical industrial process that can be optimized using artificial intelligent (AI)-based technologies like machine learning, and reinforcement learning.

REFERENCES

Abdirad M., and Krishnan K., 2021. Industry 4.0 in logistics and supply chain management: a systematic literature review, *Engineering Management Journal* 33(3), 187–201. doi: 10.1080/10429247.2020.1783935

Abraham A., Kovalev S., Tarassov V., and Snášel V., 2016. Preface, *Advances in Intelligent Systems and Computing* 451, v–vi. doi: 10.1007/978-3-319-33609-1

Arshinder A.K., Kanda A., and Deshmukh S.G., 2008. Supply chain coordination: perspectives, empirical studies and research directions, *International Journal of Production Economics* 115(2), 316–335. doi: 10.1016/j.ijpe.2008.05.011

Awad H.A.H., and Nassar M.O., 2010a. A broader view of the supply chain integration challenges, *International Journal of Innovation, Management and Technology* 1(1), 51.

Awad H.A.H., and Nassar M.O. 2010b. Supply chain integration: definition and challenges, *Proceedings of the International MultiConference of Engineers and Computer Scientists* IMECS 2010, Hong Kong, pp. 405–409.

Bag S., Yadav G., Dhamija P., and Kataria K.K., 2021. Key resources for Industry 4.0 adoption and its effect on sustainable production and circular economy: an empirical study, *Journal of Cleaner Production* 281, 125233. doi: 10.1016/j.jclepro.2020.125233

Ballou R.H., Gilbert S.M., and Mukherjee A., 2000. New managerial challenges from supply chain opportunities, *IEEE Engineering Management Review* 28(3), 7–16. doi: 10.1016/S0019-8501(99)00107-8

Bigliardi B., Bottani E., and Casella G., 2020. Enabling technologies, application areas and impact of Industry 4.0: a bibliographic analysis, *Procedia Manufacturing*, 42, 322–326.

Chen I.J., and Paulraj A. 2004. Understanding supply chain management: critical research and a theoretical framework, *International Journal of Production Research* 42(1), 131–163. doi: 10.1080/00207540310001602865

Cheng G.J., Liu L.T., Qiang X.J., and Liu Y., 2017. Industry 4.0 development and application of intelligent manufacturing, *Proceedings - 2016 International Conference on Information System and Artificial Intelligence, ISAI 2016*, pp. 407–410.

Cigánek J., Kozák S., and Kozáková A., 2018. Institute of electrical and electronics engineers, IEEE Czechoslovakia Section. Control Systems Society Chapter, and Slovenská spoločnosť pre kybernetiku a informatiku. 2018, *2018 Cybernetics & Informatics (K&I) : Proceedings of the 29th International Conference : Lazy Pod Makytou, Slovakia*.

Dalenogare L.S., Benitez G.B., Ayala N.F., and Frank A.G., 2018. The expected contribution of Industry 4.0 technologies for industrial performance, *International Journal of Production Economics* 204(July), 383–394. doi: 10.1016/j.ijpe.2018.08.019

El Hamdi S., Abouabdellah A., and Oudani M., 2019. Industry 4.0: fundamentals and main challenges, *2019 International Colloquium on Logistics and Supply Chain Management, LOGISTIQUA, Paris, France*, pp. 1–5. doi: 10.1109/LOGISTIQUA.2019.8907280

Erro-Garcés, A., 2019. Industry 4.0: defining the research agenda, *Benchmarking* 28(5), 1858–1882. doi: 10.1108/BIJ-12-2018-0444

Foidl H., and Felderer M., 2016. Research challenges of Industry 4.0 for quality management. In: Felderer M., Piazolo F., Ortner W., Brehm L., and Hof H.J., eds., *Innovation in Enterprise Information Systems Management and Engineering ERP Future 2015. Lecture Notes in Business Information Processing*. Springer, Cham, Switzerland, Vol. 245, pp. 121–137. doi: 10.1007/978-3-319-32799-0_10

Gundlach G.T., Bolumole Y.A., Eltantawy R.A., and Frankel R., 2006. The changing landscape of supply chain management, marketing channels of distribution, logistics and purchasing, *Journal of Business and Industrial Marketing* 21(7), 428–438. doi: 10.1108/08858620610708911

Hwang J.S. 2016. The Fourth Industrial Revolution (Industry 4.0): intelligent manufacturing, *SMT Prospects & Perspectives* 10–15.

Khan A., and Turowski K. 2016. A perspective on Industry 4.0: from challenges to opportunities in production systems, *IoTBD 2016 – Proceedings of the International Conference on Internet of Things and Big Data (IoTBD), Rome, Italy*, pp. 441–448. doi: 10.5220/0005929704410448

Kumar S., Raut R.D., Narwane V.S., and Narkhede B.E., 2020. Applications of Industry 4.0 to overcome the COVID-19 operational challenges, *Diabetes and Metabolic Syndrome: Clinical Research & Reviews* 14(5), 1283–1289. doi: 10.1016/j.dsx.2020.07.010

Mangla S.K., Sharma Y.K., Patil P.P., Yadav G., and Xu J., 2019. Logistics and distribution challenges to managing operations for corporate sustainability: study on leading Indian diary organizations, *Journal of Cleaner Production* 238, 117620. doi: 10.1016/j.jclepro.2019.117620

Mulky A.G., 2013. Distribution challenges and workable solutions, *IIMB Management Review* 25(3), 179–195. doi: 10.1016/j.iimb.2013.06.011

Oghazi P., Rad F.F., Karlsson S., and Haftor D., 2018. RFID and ERP systems in supply chain management, *European Journal of Management and Business Economics* 27(2), 171–182. doi: 10.1108/EJMBE-02-2018-0031

Onik M.M.H., Miraz M.H., and Kim C.S., 2018. A recruitment and human resource management technique using blockchain technology for Industry 4.0, *ArXiv* 1812.03237v1.

Pilloni V., 2018. How data will transform industrial processes: crowdsensing, crowdsourcing and big data as pillars of Industry 4.0, *Future Internet* 10(4). doi: 10.3390/fi10030024

Preuveneers D., and Ilie-Zudor E., 2017. The intelligent industry of the future: a survey on emerging trends, research challenges and opportunities in Industry 4.0, *Journal of Ambient Intelligence and Smart Environments* 9(3), 287–298. doi: 10.3233/AIS-170432

Raj A., Dwivedi G., Sharma A., Lopes de Sousa Jabbour A.B., and Rajak S., 2020. Barriers to the adoption of Industry 4.0 technologies in the manufacturing sector: an inter-country comparative perspective, *International Journal of Production Economics* 224, 107546. doi: 10.1016/j.ijpe.2019.107546

Salam M.A., 2017. The mediating role of supply chain collaboration on the relationship between technology, trust and operational performance: an empirical investigation, *Benchmarking* 24(2), 298–317. doi: 10.1108/BIJ-07-2015-0075

Sony M., 2020. Pros and cons of implementing Industry 4.0 for the organizations: a review and synthesis of evidence, *Production & Manufacturing Research* 8. doi: 10.1080/21693277.2020.1781705

Sousa S., Rodrigues N., and Nunes E., 2017. Application of SPC and quality tools for process improvement, *Procedia Manufacturing* 11(June), 1215–1222. doi: 10.1016/j.promfg.2017.07.247

Trstenjak M., 2018. Challenges of human resources management with implementation of Industry 4.0, University of Zagreb, Croatia 1–11.

Vaidya S., Ambad P., and Bhosle S., 2018. Industry 4.0 - a glimpse, *Procedia Manufacturing* 20, 233–238. doi: 10.1016/j.promfg.2018.02.034

Zhao C.M., Wei X.-P., Yu Q.-S., Deng J.-M., Cheng D.L., and Wang G.X., 2005. Photosynthetic characteristics of *Nitraria tangutorum* and *Haloxylon ammodendron* in the Ecotone between oasis and desert in Minqin, Region, Country, *Acta Ecologica Sinica* 25(8), 1908–1913.

Zheng P., Wang H., Sang Z., Zhong R.Y., Liu Y., Liu C., Mubarok K., Yu S., and Xu X., 2018. Smart manufacturing systems for Industry 4.0: conceptual framework, scenarios, and future perspectives, *Frontiers of Mechanical Engineering* 13(2), 137–150. doi: 10.1007/s11465-018-0499-5

Zhou K., Liu T., and Zhou L., 2016. Industry 4.0: towards future industrial opportunities and challenges, *12th International Conference on Fuzzy Systems and Knowledge Discovery, FSKD 2015*, Zhangjijie Shi, pp. 2147–2152.

17 Emulation of Industrial Automation Using Programmable Logic Controller

Yogesh Singh, Ritula Thakur,
and Stalin Kumar Samal

CONTENTS

17.1 INTRODUCTION

Automation is the technique to automate and expedite the industrial processes with very little or no manual intervention. It results in increased accuracy, fewer errors, less downtime due to less quality failures, increased production to achieve production planning, and less manpower required thereby curbing expenses. The impact of automation is immense and it deeply affects all kinds of industries such as (1) in the manufacturing industry computer numerical control (CNC) machines and robotized assembly lines; (2) in the financial industry high-frequency trading; (3) in shipping the autopilots in the aeronautical and space industry; (4) in hotels the ventilation and climate controls; (5) in the chemical, pharmaceutical, and semiconductors industries clean room technology and process automation; and even (6) in Internet search engines like Web crawlers.

The programmable logic controller (PLC), also known as the industrial computer, is the major component in the industrial automation sector. Even in the age of Industry 4.0 and industrial Internet it can be assumed that these controllers will continue to be required to a considerable extent (Langmann and Stiller, 2019). Due to its robust construction, exceptional functional features, like proportional-integral-derivative (PID) controllers, sequential control, timers and counters, ease of programming, reliable controlling capabilities, and ease of hardware usage, make the PLC more than a special-purpose digital computer in industries as well as in other control-system areas. Different types of PLCs from a vast number of manufacturers are available in today's market. The various functionalities of PLCs have evolved over the years to include sequential relay control, distributed control systems, process control, motion control, and networking. The PLC control system is the control key component, utilizing a special input/output (I/O) module to form the hardware of the control system with a small amount of measurement and peripheral circuit to realize control of the whole system through programming.

The population in India has been growing at the rate of 18% in the last decade (Saha et al., 2017). It is very difficult to handle the waste generated by the increased population. Solid waste management has become one of the main issues in both urban and rural areas all over the world (Johny et al., 2013). Automated waste segregation can resolve various problems that are being faced in municipal corporations such as risk of hazardous and contaminated waste management, accuracy improvement, and enhancement in speed of waste management.

The rest of the chapter is organized as follows. First, the system description is explained in two parts as hardware description and software description. Then a case study is discussed. Next, the basics of ladder logic programming and the program for our case study are explained, and finally, a conclusion is given.

17.2 SYSTEM DESCRIPTION

17.2.1 Hardware Description

Some of the manufacturers or types of PLCs include the following: Allen Bradley (AB), Asea Brown Boveri (ABB), Siemens, Mitsubishi, Hitachi, Delta, General Electric (GE), Honeywell, and Modicon. Every PLC system needs at least these three modules: central processing unit (CPU), power supply, and one or more I/O.

1. *CPU module:* It consists of a central processor and its memory. The processor is responsible for doing all the necessary computations and data processing by accepting the inputs and producing appropriate outputs. Memory includes both ROM and RAM memories.
2. *Power supply module:* These modules supply the necessary power required for the whole system by converting the available AC or DC required for CPU and I/O modules. The output 5 V DC drives the computer circuitry, and in some PLCs 24 DC on the bus rack drives a few sensors and actuators.
3. *I/O modules:* Input and output modules of the PLC allow for the sensors and actuators to be connected to the system to sense or control the real-time variables such as temperature, pressure flow, etc.

The structure of the PLC is shown in Figure 17.1.

The PETRA kit includes two conveyors and three dispensers containing an assortment of parts, some of which are outside the specification. A pneumatic pick-and-place plunger travels along a carriage gantry to place parts on one conveyor and

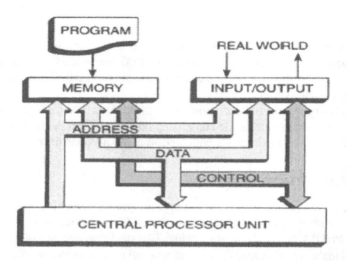

FIGURE 17.1   Structure of programmable logic controller (PLC).

remove them from one conveyor to another. A pneumatically actuated gripper arm transfers parts between the conveyors. A range of pneumatic and optoelectronic sensors allows the parts to be tested for compliance with the specification as they pass along the conveyors.

Figure 17.2 illustrates a PETRA kit with two conveyor belts, a thickness detector, a slot detector, an arm mover, a hole detector, and a cut out length detector, which has been clicked from the lab.

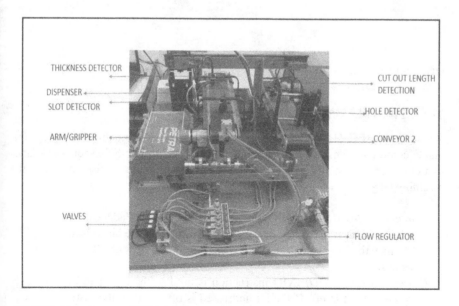

FIGURE 17.2   Image of PETRA kit.

17.2.2 Software Description

There are so many connections in industries, hence, it is very difficult to change the connection physically according to need. The PLC can be used to make this easier. It can provide flexible, rugged, and easily programmable controllers to replace hard-wired relay logic systems.

Supervisory control and data acquisition (SCADA) is used for high-level process supervisory management, while comprising other peripheral device PLCs and discrete PID controllers to interface with process plants and machinery.

17.3 CASE STUDY

In this case, we choose to develop a simple program for checking different dimensions of an object (piece of plate) and collected 20% of them for sample purposes using a PETRA kit. If all the dimensions like thickness and slot of an object are matched to the set value, then the moving arm will pick the object and place it on conveyor belt 2. On conveyor belt 2 cut out length has to be checked and if found suitable than one piece of plate will be picked up by the carriage for sample representation.

Figure 17.3 illustrates an object whose specifications have been checked.

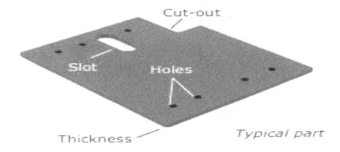

FIGURE 17.3   Image of object.

17.4 PROGRAMMING AND MODELING

According to IEC 61131-3, five programming languages are defined for programmable control systems: ladder diagram (LD), structured text (ST), sequential function chart (SFC), function block diagram (FBD), and instruction list, similar to the assembly language (IL).

Examine if closed (XIC): XIC instruction is used in the ladder to determine whether a bit is on. When instruction is executed, if the bit addressed is on (1), then the instruction is evaluated as true, and address off (0) is evaluated as false.

Examine if open (XIO): XIO instruction is used in the ladder to determine whether a bit is off. If the bit addressed is off (0), then the instruction is evaluated as true.

The output energize is used as an OTE instruction in the ladder program to turn on a bit when rung () conditions are evaluated as true.

Output latch (OTL): The OTL instruction is a retentive output instruction that can only turn on a bit. This instruction is traditionally used in pairs with an output unlatch (OTU) instruction, with both instructions addressing the same bit.

Output unlatch (OTU): The OTU instruction is a retentive output instruction that can only turn off a bit, used in pairs with an OTL instruction.

Timer: The timers are used for providing a certain amount of time delay in running the operation for a particular time. The retentive timer is also used for delay, but if in between the processes input is off it will stop. Whenever input is on counting will start from the resumed value.

Counter: The counters are used for counting physical objects passing through. There are basically two types: up (CTU) and down (CTD). CTUs count in increasing order with passes of each pulse, and whenever it reaches a set value it will stop. On the other hand, the CTD will start counting from the set value and decrease in each step by passing each pulse. The reset (RES) key is used for resetting the timer and counter.

Math instruction: Math instructions are used for performing mathematical operations like ADD for addition, SUB for subtraction, MUL for multiplication, DIV for division, TOD for integer, to BCD and SQR for finding the square root of the value.

Comparison instructions: These instructions are used for comparing two numbers like LES for less than, GRT for greater than, EQU for equal to, and LEQ for less than equal to, etc.

Program outline: In our case study we have used ladder logic programming. We divided the complete case study in two parts to make it easy to understand. In the first part of the programming we have checked the thickness and detected a hole in the object; if both are found appropriate then the moving arm has to place the object on conveyor 2. In the second part the cut out length of the object has to be observed. If the length of the cut out is as per description then it has to move through the counter, otherwise the conveyors have to stop. When five objects pass through the counter then the carriage has to collect one piece out of five for sample purposes.

Figure 17.4 shows a ladder logic program used to detect the thickness and slot of object.

In ladder logic programming, there are two rails parallel to each other and each rung is connected between these rails as shown in Figures 17.4 and 17.5. In the first rung one switch and two conveyor belts are connected. The switch shows the condition of the dispenser. If the dispenser is not empty then the switch will go on and conveyors will start to run. All items will appear as green. If the dispenser is empty then conveyors will remain stationary.

In the second rung output of two sensors, the thickness detector and slot detector act as a switch. If the thickness and slot of the object is as per set value then the timer

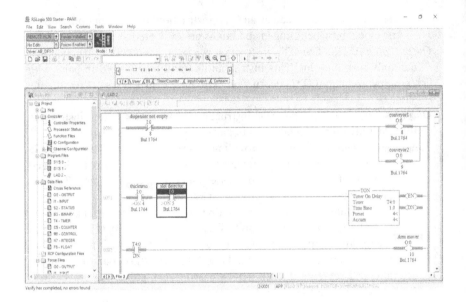

FIGURE 17.4 Ladder logic program of PLC used for detection of thickness and slot of object.

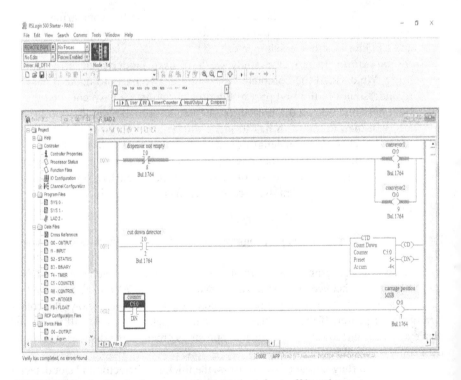

FIGURE 17.5 Ladder logic program for detection of cut-off length.

will on. The timer is used to provide a 4-second delay. This delay is the time taken by the object to reach up to the moving arm from the slot sensor by conveyor belt 1. After 4 seconds the arm mover will hold the object and place it on conveyor belt 2. This way only appropriate specification-based objects will reach the collection box through conveyor 2.

Figure 17.5 shows the ladder logic program used for detecting the cut out length of an object and counting number of the object passing through it.

In the second program, as shown in Figure 17.5, the object has been put on conveyor belt 2 where the cut out sensor has to measure the length. If the cut out is appropriate as per specification, then this piece will pass and the counter will start counting. When five pieces pass through the counter then the carriage will collect one piece and put it in a different dispenser. This way we were able to collect 20% of the samples of the completely finished objects.

17.5   CONCLUSIONS

The logic of the first rung of the program will check whether the dispenser is empty or not, If there are objects in the dispenser, only then will the conveyor belts start to run. After that thickness and slot will be detected, and if found as per set value then the moving arm will place the object on conveyor belt 2 from conveyor belt 1. Detecting a suitable object as per given specification and collecting 20% of total pieces in different dispensers has been explained in this chapter.

REFERENCES

Johny J., Joy A., Sunny D., Joseph, B.M., and Jamal S.M., 2013. Automatic plastic separating technology for solid waste disposal, *International Journal of Civil, Structural, Environmental and Infrastructure Engineering Research and Development (IJCSEIERD)* 3(2), 99–108.

Langmann R., and Stiller M., 2019. The PLC as a smart service in Industry 4.0 production systems, *Applied Sciences* 9(18), 3815. doi: 10.3390/app9183815

Lashin, M., 2014. Different applications of programmable logic controller (PLC), *International Journal of Computer Science, Engineering and Information Technology* 4(1), 27–32.

Saha H.N., Auddy S., Pal S., Kumar S., Pandey S., Singh R., Kumar Singh A., Banerjee S., Ghosh D., and Saha, S., 2017. Waste Management using Internet of Things (IoT), *2017 8th Annual Industrial Automation and Electromechanical Engineering Conference (IEMECON)*, Bangkok, Thailand.

18 A Review of Machine Learning Applications in Materials Science

Mandeep Kaur Dhami, Inderpal Pasricha, Ravreet Kaur, and Anmoldeep Singh Sidhu

CONTENTS

18.1 INTRODUCTION

Manufacturing is the process of transforming raw materials into products for societal needs. Manufacturing requires different resources such as materials, energy, capital, and people. In manufacturing, engineering design is the first step of which material design is a significant part. Material design is the selection of the required physical, chemical, and mechanical properties, ensuring the expected life of the product or its elements. Materials science, which aims to investigate the effect of the structure (electron, crystalline, micro, and macro) on materials properties is therefore the starting point in designing new materials for various applications in the near and distant future (Dobrzanski, 2006).

However, the time frame for discovering new materials from initial research to first use is remarkably long, typically approximately 10–20 years. New materials research comprises seven discrete stages: discovery, development, property optimization, system design and integration, certification, manufacturing, and deployment. Experimental measurement and computer simulation and analysis are two conventional methods that are widely adopted in the field of materials science. Experimental measurement, which usually includes microstructure and property analysis, property measurement, synthetic experiments, and so on, is an established method of materials research. Alternatively, computer simulation is another approach in which existing theory is exploited for analysis using computer programs. Materials design by using computer simulation and analysis leads to reduction in time and cost of materials development. During the last two decades, computer simulations in materials science have been employed for the discovery and design of new materials using computational results, data mining, and machine

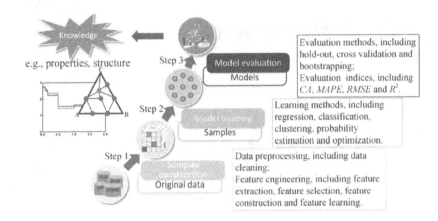

FIGURE 18.1 General process of machine learning application in materials science. (From Liu et al., 2017.)

learning (Liu et al., 2017). Figure 18.1 illustrates the general process of applying machine learning in materials science.

Machine learning is a method of data analysis that automates analytical model building. Using algorithms that iteratively learn from data, machine learning allows computers to find hidden insights without being explicitly programmed where to look. Supervised learning is by far the most widespread form of machine learning in materials science (Schmidt et al., 2019). Figure 18.2 depicts the workflow applied in supervised learning.

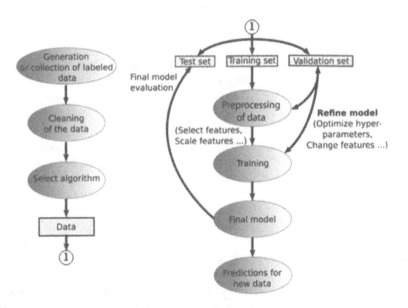

FIGURE 18.2 Supervised machine learning workflow. (From Schmidt et al., 2019.)

In the following section excerpts of applications of machine learning in materials science reported in the literature in recent years are presented in chronological order, followed by a summary of salient features of machine learning applications in materials science for different objectives.

18.2 REVIEW OF MACHINE LEARNING APPLICATIONS IN MATERIALS SCIENCE

Pilania et al. (2013) presented a machine learning-based approach for efficiently and accurately predicting a diverse set of properties of material systems, with the objective of discovering new materials with desirable properties. Charge level density and coarse-level chemo-structural descriptors were employed to map the material attributes to its properties. A family of one-dimensional chain systems was used for and presented a general formalism that allowed them to discover some decision rules. These rules established a mapping between easily accessible attributes of a system and its properties by using the kernel ridge regression algorithm of machine learning. Thus, it was shown that material fingerprints based on either chemo-structural or the electronic charge density distribution can be used to make ultrafast and accurate property predictions using machine learning.

Bélisle et al. (2015) used machine learning for predicting physical properties of materials on the basis of materials science data. Three approaches in machine learning, neural network, polynomial interpolation, and Gaussian processes, were compared to predict properties for new systems based on three sets of materials science data (i.e., molar volume, electrical conductivity, and martensite start temperature). It was demonstrated that the Gaussian process regression technique gave very good predictions for all three sets of tested data. It was also observed that a thorough knowledge of the problem beforehand is critical in selecting the most successful machine learning technique.

Yosipof et al. (2015) exhibited the use of machine learning in the development of new metal oxides for designing solar cells with higher power conversion efficiencies. A data mining workflow was developed and applied to the analysis of solar cell libraries based on titanium and copper oxides. The results demonstrated that quantitative structure-activity relationship (QSAR) models have good prediction statistics for multiple solar cells properties. The models also highlighted important factors affecting the properties similar to experimental findings.

Liu et al. (2015) presented a machine learning-based approach of material design that has an optimum microstructure with the desired properties. The microstructure of iron-gallium alloy was optimized to obtain a low elastic modulus and high magnetostrictive strains. Due to the high dimensionality of microstructure space, the multiobjective design requirement, and non-uniqueness of solutions, the traditional search-based optimization methods were observed to be incompetent; therefore, machine learning methodology was employed for this purpose. A systematic framework consisting of random data generation, feature selection, and classification algorithms was developed and applied to five design problems that involved identifying microstructures satisfying both linear and nonlinear property constraints.

Li et al. (2017) applied machine learning to search for multi-metallic copper catalysts for electrochemical reduction of carbon dioxide. The machine-learning algorithms were integrated into the descriptor-based design approach to achieve rapid screening of transition-metal catalysts. A numerical representation of surface metal atoms was engineered by using easily accessible features of local electronegativity and effective coordination dependent on the surroundings of an adsorption site, together with the intrinsic properties of active metal atoms including electronegativity, ionic potential, and electron affinity. The machine learning model was able to capture complex, nonlinear adsorbate/substrate interactions. It was observed that compared with the traditional high-throughput computational and experimental trial-and-error approach, the machine-learning chemisorption models showed great potential in accelerating the discovery of interesting catalytic materials.

Petrich et al. (2017) presented a machine learning framework for crack detection in lithium-ion battery electrodes. They proposed a classification model, using simulated data based on a three-dimensional (3D) stochastic particle model. It was used to inspect pairs of particles with possible breakages using tomographic 3D images. The test was validated by using hand-labeled data from a real electrode, achieving an overall accuracy of 73%.

Borboudakis et al. (2017) used machine learning to predict gas adsorption in metal-organic frameworks (MOFs), which belong to a rapidly growing family of hybrid inorganic-organic nanoporous materials. MOFs have shown exceptional performance in gas storage and separation of hydrogen and carbon dioxide gases. Using machine learning methods and automated analysis protocols, it was demonstrated that the chemical properties of MOFs are predictable. Therefore, the machine learning approach could be used for designing new MOFs.

Huber (2018) used artificial neural networks (ANNs) to determine the average coordination number without any knowledge of the fully connected structure, with the objective of predicting the macroscopic mechanical properties of 3D open pore materials with different topologies. This was achieved by feeding statistical information about the local coordination numbers of detectable junctions to an ANN. Finite-element (FE) beam models were used for generating data on the mechanical behavior of structures with different topologies. Data mining was used to establish relationships between topological parameters and mechanical properties. Using ANNs resulted in a reduction of the error by a factor of 2 vis-à-vis results obtained with conventional regression using Young's modulus, yield strength, and Poisson's ratio as factors and average coordination number as a response.

Snow et al. (2019) demonstrated the use of machine learning to correlate the grain boundary structure with the properties of crystalline materials. Common neighbor analysis (CAN) signatures of atomic structures were extracted using tools available in Python and MATLAB. The resultant descriptors in the form of a fraction of unique atomic environment (UAE) were used in machine learning algorithms as input with energy of the grain boundary as the response variable, for the development of an atomic structure-property model for grain boundaries. It was observed that machine learning approaches present an attractive route to develop atomic structure-property models for grain boundaries because of the complexity of their structures.

Homer et al. (2019) deployed wavelet-based convolutional neural networks to characterize the complete 3D atomic structure of a grain boundary. The average smooth overlap of atomic position (SOAP)-based representation was used as input to the machine learning algorithm, which predicted grain boundary energy, mobility, and shear coupling. It was observed that the machine learning algorithm can play an important role in grain boundary structure learning by learning well on larger data sets and providing physically interpretable information.

Steinberger et al. (2019) used a machine learning-based approach for classification of dislocation microstructures from different dislocation density field variables. The dislocation microstructure data generated using discrete dislocation dynamics code were used as input to a Gaussian naive Bayes classifier. It was observed that the accuracy of machine learning models varies with different sets of microstructure features and spatial discretization. The approach was observed to be useful for understanding the complex relation between dislocation microstructures and the emerging mechanical behavior, which is required for designing new materials with tailored material properties.

Richert et al. (2019) employed an ANN to predict ligament radii in open-pore materials. The ANN was trained by different ligament geometries with varying ligament shapes and solid fraction. The input to the ANN consisted of a vector comprising radii computed by the three algorithms, thickness (Th), Euclidean distance transformation (EDT), and smallest ellipse (SE), along with normalized position. The output of the ANN was the corrected radii value. Thus, the ANNs were applied for reconstruction of the true geometry for prediction of mechanical properties with higher accuracy.

Würger et al. (2019) used machine learning to estimate corrosion inhibition properties of untested molecules to develop new approaches that can prevent or control corrosion and degradation processes in magnesium for specific application. The approach presented combined corrosion experiments, machine learning, data mining, density functional theory calculations, and molecular dynamics to obtain a detailed atomistic understanding of the inhibition mechanisms for different additives.

Castillo and Kalidindi (2019) used machine learning for robust extraction of intrinsic material parameters from available experimental observations from spherical indentation stress strain protocols. A Bayesian inference framework was presented that enabled the specification of uncertainty in the measurement data. The framework demonstrated potential for significantly speeding up the materials characterization by focusing on experiments that are likely to deliver the maximum value in establishing the desired properties. The introduction of a Bayesian framework greatly reduced the number of simulations necessary to establish the functional dependence of the indentation elastic modulus given the lattice orientation and the intrinsic single crystal elastic stiffness parameters. The utility of the novel framework was demonstrated for a cubic polycrystalline Fe-3%Si sample and a hexagonal polycrystalline pure titanium sample.

Reimann et al. (2019) presented a new machine learning-based approach to capture the mechanical response of various microstructures under different loads. The input data consisted of different stress-strain curves obtained from micromechanical simulations of a wide range of microstructures. The micromechanical simulations are comprised of microstructure modeling by a representative volume element (RVE) and describing the anisotropic mechanical behavior of individual grains by a crystal plasticity model. The micromechanical models are subjected to mechanical loads in an FE simulation, and their macroscopic behavior is obtained by a homogenization procedure. The ML algorithm suggested grain-size distributions, grain morphologies, and crystallographic textures, which yield the desired mechanical response for a given application.

Menon et al. (2019) presented a general hierarchical machine learning model for predicting mechanical properties of thermoplastic and thermoset polyurethanes. The algorithm was trained on a library of 18 polymers using monomer chemistry, index, and chain architecture for predicting stress-at-break, strain-at-break, and Tan delta. It was observed that integration of an intermediate layer of variables comprising domain knowledge-based physicochemical factors with an input layer comprising experimentally determined variables significantly improved the model performance.

Iwasaki et al. (2019) developed a spin-driven thermoelectric material with anomalous Nernst effect by using an interpretable machine learning method called factorized asymptotic Bayesian inference hierarchical mixture of experts (FAB/HMEs). Based on prior knowledge of materials science and physics, the interpretable machine learning algorithm provided new knowledge about spin-driven thermoelectric materials. The new knowledge and correlations obtained were used to synthesize a novel spin-driven thermoelectric material with the largest thermopower.

Tawfik et al. (2019) developed a machine learning method to calculate entropy, specific heat, effective polycrystalline dielectric function, and band gap of materials, which otherwise are calculated using quantum mechanical methods with computational challenges. The materials were described mathematically using property-labeled materials fragment descriptors. These models were found to be sufficiently accurate for allowing rapid screening of large numbers of crystal structures to accelerate material discovery.

Takahashi and Takahashi (2019) proposed a machine learning-based approach for estimation of the crystal structure using material data. Two types of machine learning are implemented to understand and predict the crystal structure from a materials database. Unsupervised machine learning-based clustering was implemented to reveal the hidden data structure. Supervised machine learning-based classification was used to classify the crystal structure. The stability of predicted materials was evaluated and confirmed through the implementation of first-principles calculations. It was observed that applying machine learning to a materials database can act as a guide toward determining the crystal structure, leading to the advancement of material design.

Xiong et al. (2020) evaluated the performance of five machine learning algorithms to predict the mechanical properties of different types of steel. Nearly four hundred data sets, characterized by chemical composition, processing parameters, inclusion parameters, and mechanical properties of different types of steels were used for training of machine learning algorithms. Random forest regression showed the best performance with exhibited high validation accuracy in predicting four mechanical properties of steels, such as fatigue strength, tensile strength, fracture

TABLE 18.1

Salient Features of Machine Learning Applications in Materials Science Reported in the Literature

Reference	Objective	Modeling Data	Machine Learning Algorithms
Pilania et al. (2013); Yosipof et al. (2015); Liu et al. (2015); Huber (2018); Reimann et al. (2019); Menon et al. (2019); Xiong et al. (2020)	Mapping between crystal/atomic structure and properties	Charge level density, chemo-structural descriptors, molar volume, electrical conductivity, martensite start temperature, body centered cubic (BCC) crystal plasticity model, Young's modulus, yield strength, Poisson's ratio, average coordination number, stress-strain data, grain-size distributions, grain morphologies, crystallographic textures, structural group, chain length, molecular weight, density, Fourier transformed infrared spectroscopy (FTIR) variables, stress-at-break, strain-at-break, and Tan delta, chemical compositions, processing parameters, inclusion parameters, and mechanical properties	Kernel ridge regression algorithm, Gaussian process regression technique, Chi-square test, information gain, F-score, support vector machines (SVMs), decision tree, artificial neural networks, random forest, linear least-square, k-nearest neighbors, symbolic regression
Yosipof et al. (2015); Li et al. (2017); Würger et al. (2019); Iwasaki et al. (2019)	Development of new materials	Crystal structure, thermal properties, electrical properties, electronegativity, effective coordination, ionic potential, electron affinity, corrosion inhibition data, spin moment, orbital moment, spin polarization, spin difference, density of stated (DoS) slope of atoms	Data mining, artificial neural networks, sketch-map, support vector regression, random forest regression, least absolute shrinkage and selection operator (LASSO) regression
Snow et al. (2019); Homer et al. (2019); Richert et al. (2019); Castillo and Kalidindi (2019); Takahashi and Takahashi (2019)	Characterization of crystal/atomic structure	Common neighbor analysis (CAN) signatures, unique atomic environments (UAEs) in grain boundary, energy of grain boundary, average smooth overlap of atomic position (SOAP)-based representation, grain boundary energy, mobility, shear coupling, set of elastic constants, set of crystal orientations, atomic number, number of atoms, electronegativity atomic radius, ligament geometries, solid fraction, property-labeled materials fragment (PLMF) descriptors	Principal component analysis, linear regression, wavelet transforms, artificial neural networks, convolutional neural network, Bayesian inference, Gaussian mixture model clustering, random forest, Huber regression, artificial neural network, SVMs, relevance vector machines, XGBoost

(Continued)

TABLE 18.1 *(Continued)*
**Salient Features of Machine Learning Applications in Materials Science
Reported in the Literature**

Reference	Objective	Modeling Data	Machine Learning Algorithms
Petrich et al. (2017); Steinberger et al. (2019)	Crack detection, dislocation classification	Gap distance distribution, specific surface area, sphericity, total dislocation density, excess line density, direction line density, Kröner–Nye tensor	Artificial neural networks, Gaussian naive Bayes classifier
Borboudakis et al. (2017)	Gas adsorption prediction in metal-organic frameworks	Structural parameters like organic linker, metal cluster, and functional groups	Support vector regression, random forest, ridge linear regression

strength, and hardness. Symbolic regression gave explicit mathematical expressions of the four mechanical properties as functions of the four important features, tempering temperature and the percentage of C, Cr, and Mo, revealing the required features of a novel antifatigue high-strength steel.

18.3 SUMMARY

The salient features of machine learning applications in materials science reported in the literature and presented in the previous section are summarized in terms of the objectives, modeling data, and machine learning algorithms in Table 18.1.

Machine learning has been employed by researchers in materials science for a variety of applications such as designing new materials with required properties, establishing structure-property relationships, characterization of crystal structure at various scales, classification of dislocation microstructure, etc. The data sets used for modeling and training of machine learning algorithms in materials science are comprised of atomic data, microstructure data, thermal, electric, physical, chemical, and mechanical properties. These data sets are typically very large in size. Therefore, to extract useful information and inference from experimental and simulation data, various machine learning algorithms related to clustering, regression, and classification have been applied in the field of materials science. Python programming language is the preferred choice for implementing machine learning algorithms in materials science, as it is an open source software and a variety of open source libraries for materials science are also available for Python. Machine learning algorithms have either performed better than, or refined the output of, conventional computational models used for different tasks ranging from electronic structure calculations to continuum macroscopic analysis. Thus, it may be stated that machine learning will continue to be applied extensively in materials science for extracting new knowledge as well as speeding up fundamental and applied research.

REFERENCES

Bélisle E., Huang Z., Le Digabel S., and Gheribi A.E., 2015. Evaluation of machine learning interpolation techniques for prediction of physical properties, *Computational Materials Science* 98, 170–177.

Borboudakis G., Stergiannakos T., Frysali M., Klontzas E., Tsamardinos I., and Froudakis G.E., 2017. Chemically intuited, large-scale screening of MOFs by machine learning techniques, *NPJ Computational Materials* 3, 40. doi:10.1038/s41524-017-0045-8

Castillo A., and Kalidindi S.R., 2019. A Bayesian framework for the estimation of the single crystal elastic parameters from spherical indentation stress-strain measurements, *Frontiers in Materials* 6, 136. doi:10.3389/fmats.2019.00136

Dobrzanski L.A., 2006. Significance of materials science for the future development of societies, *Journal of Materials Processing Technology* 175, 133–148.

Homer, E.R., Hensley D.M., Rosenbrock C.W., Nguyen A.H., and Hart G.L.W., 2019. Machine-learning informed representations for grain boundary structures, *Frontiers in Materials* 6, 168. doi: 10.3389/fmats.2019.00168

Huber N., 2018. Connections between topology and macroscopic mechanical properties of three-dimensional open-pore materials, *Frontiers in Materials* 5, 69. doi: 10.3389/fmats.2018.00069

Iwasaki Y., Sawada R., Stanev V., Ishida M., Kirihara A., Omori Y., Someya H., Takeuchi I., Saitoh E., and Yorozu S., 2019. Identification of advanced spin-driven thermoelectric materials via interpretable machine learning, *NPJ Computational Materials* 5, 103. doi: 10.1038/s41524-019-0241-9

Li Z., Ma X., and Xin H., 2017. Feature engineering of machine-learning chemisorption models for catalyst design, *Catalysis Today* 280(2), 232–238.

Liu R., Kumar A., Chen Z.-Z., Agrawal A., Sundararaghavan V., and Choudhary A., 2015. A predictive machine learning approach for microstructure optimization and materials design, *Scientific Reports* 5, 11551. doi: 10.1038/srep11551

Liu Y., Zhao T., Ju W., and Shi S., 2017. Materials discovery and design using machine learning, *Journal of Materiomics* 3, 159–177.

Menon A., Thompson-Colón J.A., and Washburn N.R., 2019. Hierarchical machine learning model for mechanical property predictions of polyurethane elastomers from small datasets, *Frontiers in Materials* 6, 87. doi: 10.3389/fmats.2019.00087

Petrich L., Westhoff D., Feinauer J., Finegan D.P., Daemi S.R., Shearing P.R., and Schmidt V., 2017. Crack detection in lithium-ion cells using machine learning, *Computational Materials Science* 136, 297–305.

Pilania G., Wang C., Jiang X., Rajasekaran S., and Ramprasad R., 2013. Accelerating materials property predictions using machine learning, *Scientific Reports* 3, 2810. doi: 10.1038/srep02810

Reimann D., Nidadavolu K., ul Hassan H., Vajragupta N., Glasmachers T., Junker P., and Hartmaier A., 2019. Modeling macroscopic material behavior with machine learning algorithms trained by micromechanical simulations, *Frontiers in Materials* 6, 181. doi: 10.3389/fmats.2019.00181

Richert C., Odermatt A., and Huber N., 2019. Computation of thickness and mechanical properties of interconnected structures: Accuracy, deviations, and approaches for correction, *Frontiers in Materials* 6, 327. doi: 10.3389/fmats.2019.00327

Schmidt J., Marques M.R.G., Botti S., and Marques M.A.L., 2019. Recent advances and applications of machine learning in solid state materials science, *NPJ Computational Materials* 5, 83. doi: https://doi.org/10.1038/s41524-019-0221-0

Snow B.D., Doty D.D., and Johnson O.K., 2019. A simple approach to atomic structure characterization for machine learning of grain boundary structure-property models, *Frontiers in Materials* 6, 120. doi: 10.3389/fmats.2019.00120

Steinberger D., Song H., and Sandfeld S., 2019. Machine learning-based classification of dislocation microstructures, *Frontiers in Materials* 6:141. doi: 10.3389/fmats.2019.00141

Takahashi K., and Takahashi L., 2019. Creating machine learning-driven material recipes
based on crystal structure, *Journal of Physical Chemistry Letters* 10, 283–288.

Tawfik S.A., Isayev O., Spencer M.J.S., and Winkler D.A., 2019. Predicting thermal properties of crystals using machine learning, *Advanced Theory and Simulations* 3(2),
1900208. doi: 10.1002/adts.201900208

Würger T., Feiler C., Musil F., Feldbauer G.B.V., Höche D., Lamaka S.V., Zheludkevich
M.L., and Meißner R.H., 2019. Data science based Mg corrosion engineering, *Frontiers
in Materials* 6, 53. doi:10.3389/fmats.2019.00053

Xiong J., Zhang T.-Y., and Shi S.-Q., 2020. Machine learning of mechanical properties of
steels, *Science China Technological Sciences* 63(7), 1247–1255.

Yosipof A., Nahum O.E., Anderson A.Y., Barad H.N., Zaban A., and Senderowitz H., 2015.
Data mining and machine learning tools for combinatorial material science of all-oxide
photovoltaic cells, *Molecular Informatics* 34(6–7), 367–379.

19 RCM Implementation in Electric Power Generation Unit

Navneet Singh Bhangu, G. L. Pahuja,
Rupinder Singh, and Smruti Ranjan Pradhan

CONTENTS

19.1 INTRODUCTION: BACKGROUND AND DRIVING FORCES

In the past four decades there has been a huge increase in volume, variety, and complexity of plant machinery in electric power generation units (Tsao, 2013). The views on maintenance are changing and new maintenance techniques are desired to avoid false starts, dead ends, and major breakdowns (Edwards et al., 2000). As far as the generating plants in the power sector are concerned, the irregular increase of load has led to an increase in forced outages. To make provisions for a higher degree of availability of generating plants, an improved maintenance plan needs to be introduced (Cigolini et al., 2008; Liyanage, 2010). The reliability centered maintenance (RCM) maintenance technique is based on systematic evaluation to optimize maintenance activities (Zhang and Nakamura, 2005; Bloom, 2006; Kumar et al., 2009). Initially RCM was developed for the aviation sector (Srivastava, 2005), but now it has become one of the practices of maintenance management for various industrial/ service organizations across the world (Moubray, 1997; Flapper et al., 2010). As per the reported literature, RCM maintenance techniques can bring a 30%–70% reduction in downtime and cost of maintenance (Crocker and Kumar, 2000; Smith and Hinchcliffe, 2004). Regarding the field implementation of RCM, a number of case

DOI: 10.1201/9781003203681-22

studies have been reported on electric distribution systems, railways, nuclear power plants, steel plants, hydro power plants, transmission systems, wind turbines in wind power plants, and cone crusher and oil refineries (Beehler, 1997; Goodfellow, 2000; Deshpande and Modak, 2002; Carreteroa et al., 2003; Li and Korczynski, 2004; Bevilacqua et al., 2005; Santosh et al., 2007; Heo et al., 2011; Fischer et al., 2012; Tsutsui and Takata, 2012; Sinha and Mukhopadhyay, 2015). For successful implementation of RCM, it has been widely expressed that the maintenance program must combine technical requirements with management strategy (Martorell et al., 2008; Niu et al., 2010). Many investigators have highlighted improvement in the maintenance plan in various sectors using RCM, but a lot less work has been reported on the application of RCM in electric power generation units and exploring its economic benefits. So, this persuaded us to investigate RCM as a method of working in thermal power plants. RCM has seven criteria, which may be grouped into three structured steps/phases: (1) define, (2) analyze, and (3) act. The work presented has been designed according to these three structured steps. Justifying the define phase, investigations in context of understanding the functions of each component at a thermal power plant, collection of the outage review report for critical fault identification, and availability analysis are presented. Pareto analysis and failure mode and effects analysis (FMEA) has been performed to validate the analysis phase. The act phase has been justified by presenting RCM decision worksheets and performance was evaluated thereafter. The RCM process includes systematic procedures (Billinton and Allan, 1996; Guevara and Souza, 2008) that have been adopted for a given thermal power plant (Table 19.1).

By systematically applying the RCM methodology on the most critical component of the thermal plant, an improvement in availability and revenue generation has been noted.

TABLE 19.1
RCM Procedures and Methodology

Procedure Details	Operation Level	Data Requirement	Results	
Failure data collection	System	Component failure data	Failure rate	Define
Analysis for availability	System/subsystem	Component data for calculation	Availability indices	Define
Pareto analysis	Component	Failure time	Critical components singled out	Define
Analysis of failure modes	Component	Cause of failure	Mode and reason of failure	Analyze
Estimation of effect of failure	Component	Outage review report	Composite failure frequency & effect	Analyze
Drafting preventive maintenance policy	Component	Failure modes and effects	RCM plan	Act
Predicting improvement in uptime	Component/system	Failure frequency and costs	Enhanced availability and cost benefits	Act

19.2 SELECTION OF THE SYSTEM: DEFINING STAGE

The load conditions in electric power units have recently become severe due to varying loads and there is trend of material damage (Shimomura and Sakurai, 2002; Selvik and Aven, 2011). This has caused an adverse effect on the reliability of electric power plants. Thus, the electric utility industry is supposed to maintain a high level of system availability and reliability. Keeping this in mind, a thermal power plant (GNDTP, Bathinda, India) situated in eastern Punjab of the Indian subcontinent has been chosen for investigations. The plant was established in 1969 and is made up of four units having a total capacity of 440 MW. Turbines used are the impulse type with 3000 rpm and generators are the 3-phase synchronous type, 110 MW, 11 kV with hydrogen cooling, and a DC/static exciter.

19.2.1 COLLECTION OF FAILURE DATA AND AVAILABILITY ANALYSIS

Considering the outage hours and number of failures of the plant the parameter, "availability" has been calculated to find out the annual generation loss. The availability function may be computed by using the Markov model (Balagurusamy, 2003).

$$\text{Availability}(A) = \frac{\dfrac{1}{\lambda}}{\dfrac{1}{\lambda} + \dfrac{1}{\lambda}} \tag{19.1}$$

Note: $1/\lambda$: MTBF, $1/\mu$: MTTR

Equation (19.1) may be expressed as:

$$A = \frac{\text{MTBF}}{\text{MTBF} + \text{MTTR}} \tag{19.2}$$

Note:

MTBF: mean time between failures
MTTR: mean time to repair
λ: failure rate
μ: repair rate
t: period of failure (hours)

Table 19.2 shows the plant parameters for one year (2010–2011).

The annual forced outages are responsible for unavailability of the plant due to which it suffers huge losses in annual generation and revenue. If the number and period of forced outages can be reduced, then the plant will be available for longer. This will, no doubt, enhance system uptime and more revenue can be earned by generating more electrical energy.

Thus, the preliminary survey points out that there is strong need for detailed investigations. The first step toward this is to collect data for each unit for four consecutive years and analyze for critical fault identification.

TABLE 19.2
Plant Parameters for F.Y 2010–2011

Parameters	Observed Value
No. of failures per year	43
Total outages	4239 (hr/year)
MTTR	98.58
MTBF	105.14
Unavailability	0.48
A	0.52
Annual power generation loss	1850×10^6 kWh
Outage cost/year (by considering @Rs. 3.19 INR/unit)	107×10^6 USD (Rs. 590.15 crore INR)

19.3 CRITICAL FAULT AND FUNCTIONAL FAILURE IDENTIFICATION: ANALYSIS STAGE

19.3.1 Pareto Analysis

The downtime data has been collected from the thermal power plant for the period F.Y 2007–11. The reasons for tripping were sorted out and arranged in descending order as per input structure for Pareto analysis (Bhangu et al., 2015). In the present case study, vital Pareto analysis-based components have been separated from the trivial components, many which are prone to failures. Based on the Pareto principle (Chen and Wang, 2010; Oke et al., 2008), Figure 19.1 shows a Pareto chart. As observed in Figure 19.1, first nine reasons for tripping contribute to 87% of the total outage time.

Critical ranking has been assigned to these components based on their contribution to downtime as per Table 19.3.

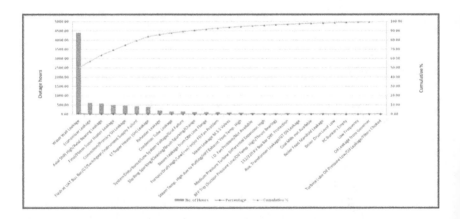

FIGURE 19.1 Pareto chart for plant downtime (based on four years data).

TABLE 19.3
Critical Faults Identified for the Plant

Component Prone to Faults	Functional Failure	Ranking as Per Criticality
Water wall tubes	Steam leakage	1
Economizer	Water leakage	2
Turbine	Axial shift high/bearing leakage	3
Final/platen super heater	Steam leakage	4
Primary/convection super heater	Steam leakage	5
Unit auxiliary transformer/switchgear	Flash	6
Low temp. super heater	Steam leakage	7
Reheater	Steam leakage	8
Condenser tube	Water leakage	9

19.3.2 FMEA

The second criterion for the analysis stage (i.e., initial part of the RCM process perform FMEA for the most critical component of the system; Bevilacqua et al., 2000). After this structural decision logic is applied to the FMEA outputs. Figure 19.2 depicts the process of FMEA. Conducting FMEA on existing processes, subsystems and components yield benefits as far as safety hazards and product-forced outages or malfunctions are concerned (Estorlio and Posso, 2010). FMEA has been performed on the most critical component of the thermal plant, i.e., the water wall tube to analyze modes and the effects of failures (Figure 19.2).

> *Function (F):* The FMEA starts with defining the function of each chosen subsystem or component. The task of water wall tubes in the boiler is to contain water for heating and converting it to steam (Figure 19.3).
>
> *Failure mode (FM):* Failure mode is the way components might fail. The cracks in the water wall tube or tube rupture have been identified as the

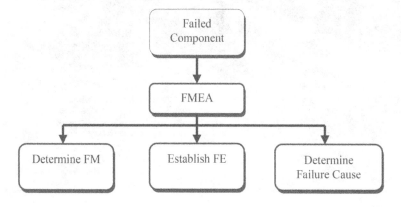

FIGURE 19.2 Process of FMEA.

FIGURE 19.3   Arrangement of water wall tubes at the site.

failure mode in this case. Figures 19.4 and 19.5 show tube rupture and tube cracks, respectively.

Failure cause (FC): The next step in the analysis process is to identify all kinds of reasonable causes that make a functional failure happen. The potential causes of failure mode in water wall tubes have been identified.

Failure effect (FE): This is the result of the failure mode on the process function. It describes what happens when each failure mode occurs. The ultimate effect of cracks in tube and tube rupture is determined to be steam leakage and ultimately forced shutdown of the unit.

Based on this information, the RCM sheet has been prepared in the operating context of the component.

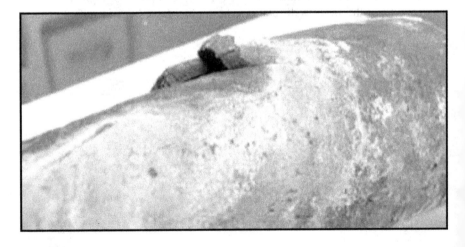

FIGURE 19.4   Tube rupture.

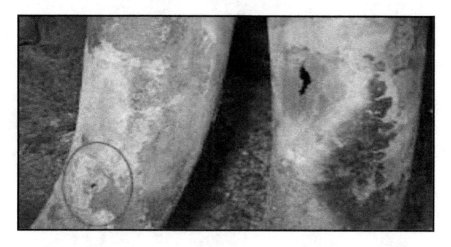

FIGURE 19.5 Tube cracks.

19.4 RCM DECISION LOGICS: ACTION STAGE

The RCM decision logic scheme has been used for evaluating the maintenance requirements (in the form of failure consequences) for significant components. The RCM information worksheet was used as input for a decision logic scheme and the output was tabulated as a decision worksheet. Commercial RCM++ software by ReliaSoft® (2011) was used for RCM analysis as per SAE JA1012 standards (Society of Automotive Engineers, 2002). The procedural steps of using RCM++ include (1) equipment selection, (2) adding function and description of equipment, (3) mentioning functional failure and failure effects and categorization of failure effects, (4) adding failure mode, and (5) ultimately selecting maintenance task. The software is helpful to identify any possible discrepancies and omissions in each FMEA-based RCM analysis. Table 19.4 shows the report generated by RCM++ for water wall tubes.

The FM-1 is a tube rupture/crack that is caused by boiler purified water becoming acidic at the condenser. The proposed task for this has been suggested to be use of equipment (gas detector) for detecting air and water leakage at the condenser. This will detect even minute leakage, and if it is repaired before growing into gigantic fault (as preventive measure), it will avoid a long downtime. Presently no such task is used, and it has been proposed to use it three times a year. A shutdown of 5–6 days is required for this one-time preventive measure, but ultimately there is going to be a net benefit of working days and uptime efficiency.

The proposed task for FM-2 is checking for cracks in the tube using ultrasonic testing instrument. This is a simple procedure that sends ultrasound into the pipe and the reflecting frequency is measured. A crack, if there is any, will be detected by this. As the mode of this failure is continuous, exposure of pipes to flame can only be evaded by this timely preventive measure. Presently no such task is prevailing at the site, so it is proposed in a similar manner as for FM-1.

TABLE 19.4
RCM Report for Water Wall Tubes

Document Number: 1/WWT

FMEA Type: Process

Prepared By: NSB

Function	Functional Failure	Failure Mode	Effect	O	FEC	Proposed Task Description	Type	Assigned Interval	Assigned Interval Units	Proposed Interval	Proposed Interval Units	Deptt. Name
1 - Water Wall Tubes												
To contain water for heating and converting to steam	Cannot maintain steam pressure at all	Tube rupture due to contaminated (acidic) boiler water by mixing of it with condenser water.	No required pressure build up of steam, hence tripping.	Y	EO	Regular Check up for condenser tube leakage using Helium gas detector or SF6 tracer gas technology for both air and water leakage	S	No such technique is used.	Nil	16	Week	Maint./ Mech.
		Tube wear down due to over exposure to flame i.e. erosion	Tripping of the plant due to decrease of steam pressure	Y	EO	Check up for tube cracks using ultrasonic probe testing or acoustic instruments	S	No such technique is used.	Nil	16	Week	Elect./ Maint.

(Continued)

TABLE 19.4 (Continued)
RCM Report for Water Wall Tubes

Document Number: 1/WWT

FMEA Type: Process

Prepared By: NSB

Function	Functional Failure	Failure Mode	Effect	O	FEC	Proposed Task Description	Type	Assigned Interval	Assigned Interval Units	Proposed Interval	Proposed Interval Units	Proposed Deptt. Name
	Maintains less pressure than required	Minute cracks in tubes due to sticking of mineral contents to tubes from coal ash by burning poor quality of coal i.e. clinker formation	Specified steam pressure not available and running of plant in such condition may cause serious problem further leading to forced longer shut down.	Y	EO	Regular poking of hopper for clinker removal.	OC	48	Weeks	16	Week	Maint.

Remarks: FEC: Failure Effect Categorization; EO: Evident Operational; O: Operation Affected; Y: Yes; S: Service; OC: On Condition; NE: Not Evident; N: No

The FM-1 for FE–B is microscopic cracks/holes in the tube due to clinker sticking. The clinker is formed mainly due to burning of poor-quality coal with greater mineral content. The root cause of this problem is unavoidable, as it is related to the quality of coal used and improving quality has many other nontechnical restrictions found at the plant under study. The preventive measure suggested is regular poking of the hopper (for clinker removal). This is required to be performed more frequently (compared with the existing practice) along with the previously mentioned tasks.

The cost factor has also been kept in mind during planning of the RCM worksheets. The tasks proposed mostly do not incur extra expenditure. The existing maintenance personnel of the respective departments will perform the tasks. The only one-time cost involved is purchasing new equipment required for testing, but this will shortly be paid back by the extra revenue earned with more uptime.

19.5   RCM AUDIT TO EVALUATE ECONOMIC BENEFITS

The mentioned RCM decision logic has been recommended to the maintenance department for execution as a pilot project, which enabled the plant organization to gain firsthand experience implementing RCM. According to this policy, the three preventive maintenance tasks have been applied on water wall tubes out of which the two tasks are service type and one is an on-condition type. The regular checks for air and water leakage in the condenser have been performed. Ultrasonic probe testing of the water wall tube for leakage has also been done. Regular poking of the hopper has been done for clinker removal as per the prescribed schedule. The preventive repairs have been performed if any "weak spot" or "loop hole" has been observed. This helped prevent major faults in the future.

The failure data of the water wall tube from unit IV was collected again from April 2011 to March 2012. The number of faults has reduced to 17 compared with 24 the previous year pre-RCM. Similarly, outage hours against this fault have reduced to 1325.75 post-RCM compared with 1894.93 in the previous year. Table 19.5 and Figure 19.3 show the comparison between the outage hours and the number of leakage faults of the water wall tube pre- and post-RCM.

With the decrease in the number of faults from 24 to 17, the outage hours have reduced by 569.18 hours. The shutdown required for performing off-line (service type) preventive maintenance (PM) procedures suggested in the RCM decision

TABLE 19.5
Comparison of Outage Hours and Number of Faults for Water Wall Tube Leakage

Sr. No.	Year	Number of faults	Outage hours
1	2010–2011	24	1894.93[a]
2	2011–2012	17	1325.75

[a] From unit IV data.

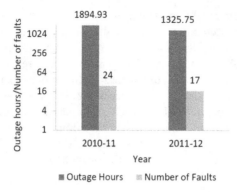

FIGURE 19.6 Comparison pre- and post-RCM.

worksheet is approximately 20 days (480 hours) per year. The scheduled or corrective maintenance was previously done for approximately 30 days. As after RCM, the suspect areas have already been tackled during PM, so it was finished well before time in 16 days, saving 14 days (336 hours). The reduced outage hours and saved maintenance hours are the total hours for which the plant has been made available and, deducting PM hours, the net availability of the unit has been found to be increased by 425.18 hours in the year 2011–2012. The extra revenue earned with this enhanced availability is Rs. 14.92 crore (USD 2.7 million) per annum. Figure 19.6 is a comparison chart of outages before and after RCM implementation.

19.6   RESULTS AND DISCUSSION

The outcome can be summarized as tabulated in Table 19.6. By introducing the RCM approach for the water wall tube of the boiler in unit IV, the numbers of faults and outage hours have reduced considerably compared with the last year. The percentage reduction in outage hours of the water wall tube is 30.04%.

Managing implementation of RCM means a change in procedure and schedule of maintenance. Convinced by the methodology and especially its economic benefits,

TABLE 19.6
Summary of Results

Sr. No.	Description	Pre-RCM	Post-RCM
1	Number of faults	24	17
2	Outage hours (hours)	1894.93	1325.75
3	Preventive maintenance days	–	20
4	Corrective maintenance days	30	16
5	Availability increase (hours)	–	425.18
6	Extra generated units (kWh)	–	$110 \times 10^3 \times 425.18$
7	Extra revenue earned (@ Rs. 3.19 INR/kWh)	–	USD 2.7 million (Rs. 14.92 crore)

there was no problem accepting the new preventive maintenance regime among the employees. The jobs were allocated as per the schedule given in the decision worksheets without any hitch.

Increased availability of the unit has generated extra revenue and some of the cost incurred (purchasing some equipment for PM) in implementing the new RCM plan will be adjusted therein. As per prescribed maintenance tasks in the RCM plan for the water wall tube, the PM has been performed for a year. One of the tasks includes condenser maintenance so that purified boiler water does not get mixed in with brackish cooling water. This PM does not only prevent water wall tube leakage but is beneficial for other components of the steam flow cycle.

19.7 CONCLUSIONS

1. The preliminary survey emphasized a strong need of failure data exploration and identification of critical faults.
2. In this case study the Pareto analyses of each unit separately for 4 years did not reach a solid conclusion, thus, a cumulative Pareto analysis has been done. The results of the study outlined eight components responsible for 87% of the outage time.
3. Further, based on FMEA, preventive tasks have been planned. The suggestions from experienced maintenance personnel acted as a backbone for the decision worksheets. An RCM preventive maintenance plan generated by RCM++ software has been proposed. The execution has confirmed reduced outages and enhanced availability of the water wall tubes (for one unit with increase in availability). The result of the study has highlighted a generic model for an RCM-based maintenance plan for thermal power plants.

This case study can be termed as the "applied RCM study" as preventive measures suggested can be generalized to only those units/plants with almost the same aging and same failure trends. Scrutinizing of the total available data can also be done by using better tools/software instead of critical fault identification by Pareto analysis. As far as future paths are concerned, all the plants in the region can be considered for this maintenance plan, keeping the plant age in consideration.

REFERENCES

Balagurusamy E. 2003. *Reliability Engineering.* Tata McGraw-Hill, New Delhi, pp. 119–120.

Beehler M.E., 1997. Reliability centered maintenance for transmission systems, *IEEE Transactions on Power Delivery*, 12(2), 1023–1028.

Bevilacqua M., Braglia M., and Gabbrielli R., 2000. Monte Carlo simulation approach for a modified FMECA in a power plant, *Quality and Reliability Engineering International* 16, 313–324.

Bevilacqua M., Ciarapica F.E., Giacchetta G., and Bertolini M., 2005. An application of BPR and RCM methods to an oil refinery turnaround process, *Production Planning & Control: The Management of Operations* 16, 716–732.

Bhangu N.S., Pahuja G.L., and Singh R., 2015. Application of fault tree analysis for evaluating reliability and risk assessment of a thermal power plant, *Energy Sources, Part A: Recovery, Utilization, and Environmental Effects* 37(18), 2004–2012.

Billinton R., and Allan R.N., 1996. *Reliability Evaluation of Power Systems*, 2nd ed. Plenum Press, New York, pp. 355–363.

Bloom N.B., 2006. *Reliability Centered Maintenance; Implementation Made Simple*, 1st ed. McGraw-Hill, Inc., New York, p. 280.

Carreteroa J., Pereza J.M., Garcıa-Carballeiraa F. et al., 2003. Applying RCM in large scale systems: A case study with railway networks, *Reliability Engineering and System Safety* 82, 257–273.

Chen Y.M., and Wang W.-S., 2010. Environmentally constrained economic dispatch using Pareto archive particle swarm optimization, *International Journal of Systems Science* 41, 593–605.

Cigolini R., Fedele L., Garetti M., and Macchi M., 2008. Recent advances in maintenance and facility management, *Production Planning & Control: The Management of Operations* 19, 279–286.

Crocker J., and Kumar U.D., 2000. Age-related maintenance versus reliability centered maintenance: A case study on aero-engines, *Reliability Engineering and System Safety* 67, 113–118.

Deshpande V.S., and Modak J.P., 2002. Application of RCM for safety considerations in a steel plant, *Reliability Engineering and System Safety* 78, 325–334.

Edwards D.J., Holt G.D., and Harris F.C., 2000. A model for predicting plant maintenance costs, *Construction Management and Economics* 18, 65–75.

Estorlio C., and Posso R.K., 2010. The reduction of irregularities in the use of process FMEA, *International Journal of Reliability Management* 27, 24–27.

Fischer K., Besnard F., and Bertling L., 2012. Reliability-centered maintenance for wind turbines based on statistical analysis and practical experience, *IEEE Transactions on Energy Conversion* 27, 184–195.

Flapper S.D.P., Fransoo J.C., Broekmeulen R., and Inderfurth K., 2010. Planning and control of rework in the process industries: a review, *Production Planning & Control: The Management of Operations* 13, 26–34.

Goodfellow J.W., 2000. Applying reliability centered maintenance (RCM) to overhead electric utility distribution systems, *Proceedings of 2000 Power Engineering Society Summer Meeting*, Seattle, WA, 566–569.

Guevara F.J., and Souza G.M., 2008. RCM Application for availability improvement of gas turbine used in combined cycle power stations, *IEEE Latin America Transactions* 6(5), 401–407.

Heo J.H., Kim M.K., Park G.P. et al., 2011. A reliability-centered approach to an optimal maintenance strategy in transmission systems using a genetic algorithm, *IEEE Transactions on Power Delivery* 26, 2171–2179.

Kumar H., Singh R., Bhangu N.S., and Pahuja G.L., 2009. Implementation of reliability centered maintenance in lamp manufacturing unit. *Journal of Manufacturing Engineering* 4, 212–219.

Li W., and Korczynski J., 2004. A reliability-based approach to transmission maintenance planning and its application in BC hydro system, *IEEE Transactions on Power Delivery* 19, 303–308.

Liyanage J.P., 2010. Maintenance for industrial systems, *Production Planning & Control: The Management of Operations* 21, 796–797.

Martorell S., Sanchez A., Villamizar M. et al., 2008. Maintenance modeling and optimization integrating strategies and human resources: theory and case study, *Proceedings of the Institution of Mechanical Engineers, Part O: Journal of Risk and Reliability* 222(3), 347–357.

Moubray J., 1997. *Reliability Centered Maintenance*. Industrial Press, Inc., New York, pp. 16–17. ISBN 0-8311-3146-2

Niu G., Yang B.-S., and Pecht M., 2010. Development of an optimized condition-based maintenance system by data fusion and reliability-centered maintenance, *Reliability Engineering and System Safety* 95, 786–796.

Oke S.A., Ofiabulu C.E., Banjo A.A., et al., 2008. The combined application of quality function development and Pareto analysis for hotel services improvement, *International Journal of Productivity and Quality Management* 3, 241–262.

ReliaSoft®. 2011. Training Guide, RCM++ Version 5. ReliaSoft Publishing, Tucson, Arizona.

Santosh T.V., Vinod G., Saraf R.K. et al., 2007. Application of artificial neural networks to nuclear power plant transient diagnosis, *Reliability Engineering and System Safety* 92, 1468–1472.

Selvik J.T., and Aven T., 2011. A framework for reliability and risk centered maintenance, *Reliability Engineering and System Safety* 96, 324–331.

Shimomura K., and Sakurai S., 2002. Advanced technologies of preventive maintenance for thermal power plants, *Hitachi Review* 51, 137–142.

Sinha R.S., and Mukhopadhyay A.K., 2015. Reliability centered maintenance of cone crusher: a case study, *International Journal of System Assurance Engineering and Management* 6, 32–35. doi: 10.1007/s13198-014-0240-7

Smith A., and Hinchcliffe G.R., 2004. *RCM – Gateway to World Class Maintenance*. Elsevier Inc., Burlington MA, p. 311.

Society of Automotive Engineers, 2002. *SAE JA1012. A Guide to the Reliability-Centered Maintenance (RCM) Standard*. Society of Automotive Engineers, Warrendale, Pennsylvania.

Srivastava S.K., 2005. *Industrial Maintenance Management*. S. Chand & Co. Ltd., New Delhi, pp. 42–43.

Tsao Y.-C., 2013. Combined production-maintenance decisions in situations with process deterioration, *International Journal of Systems Science* 44, 1692–1700.

Tsutsui M., and Takata S., 2012. Life cycle maintenance planning method in consideration of operation and maintenance integration, *Production Planning & Control: The Management of Operations* 23, 183–193.

Zhang T., and Nakamura M., 2005. Reliability based optimal maintenance scheduling by considering maintenance effect to reduce cost, *Quality and Reliability Engineering International* 21, 203–220.

20 MODWT-based Novel Health Indicator for Incipient Gear Fault Diagnosis

Mansi, Kanika Saini, Vanraj, and Sukhdeep S. Dhami

CONTENTS

20.1 INTRODUCTION

The majority of industrial machines have rotating elements such as gears, bearings, shafts, etc. Condition monitoring and fault diagnosis of rotating elements help in preventing catastrophic accidents and guaranteeing machine availability. Monitoring and analysis of machine health signature signals, such as vibration and acoustics, play a vital role in condition monitoring and fault diagnosis. However, such signals contain unwanted noise along with useful information. Therefore, the denoising of signals is essential before analyzing the signal for condition monitoring. Early fault diagnosis in low-speed machines has attracted increasing attention from both industry and academia (Heng et al., 2009; Lee et al., 2014; Vanraj et al., 2018a,b).

DOI: 10.1201/9781003203681-23

189

Gears play a very important role in machinery as they are used to transmit power from one shaft to another. Therefore, condition monitoring of gears is essential so that the productivity and efficiency of the system are not affected. The most common faults of gears are root crack, chipped tooth, wear, etc. It is necessary to monitor the gear health periodically for detecting the faults at an incipient stage. Vibration and acoustic-based fault detection techniques are used in many industries. Vibration analysis is essential in the detection of faults in gears (Randall, 2011). Vibration monitoring plays a vital role in detecting any change in the vibration signal due to gear degradation and gives an early warning. Rotating machines, such as gearing systems of mine excavators (Bartelmus and Zimroz, 2009a,b; Zimroz and Bartkowiak, 2013; Vanraj et al., 2017a,b), helicopters (Samuel and Pines, 2005, 2009), bearing systems of high speed trains (He and Wang, 2012; Lu et al., 2014), and wind turbines (De Azevedo et al., 2016) work in regimes that continuously change, which increases the chance of failures.

The vibration signals from the machines are obtained in the time domain. Along with useful information, the signals contain noise that is not required. Therefore, the denoising of signals is necessary. Denoising helps to filter the noise components and also retain details of original signals. The majority of signals in the real-world scenario are non-stationary. Denoising can be done by various techniques such as wavelet transform (WT) (Peng and Chu, 2004; Yan et al., 2014), empirical mode decomposition (EMD) (Lei et al., 2013; Vanraj et al., 2017a,b), spectral kurtosis (Antoni and Randall, 2006), singular value decomposition (SVD), minimum entropy deconvolution (Endo and Randall, 2007; Jia et al., 2017; Sawalhi et al., 2007), and stochastic resonance (Tan et al., 2009; He and Wang, 2012; Lei et al., 2013). Industrial settings, where plenty of data are gathered on a continuous basis, are generally influenced by a great deal of interference and noise, thus making noise removal an essential tool depending on whether it is Gaussian white noise or non-Gaussian noise (Shao and Nezo, 2005; Laha, 2017; Vanraj et al., 2018a,b; Wang et al., 2019).

Wavelet is considered to be an effective tool for denoising in signal processing because of its adaptability, diversity of wavelet basis, and fast computational algorithms. WT is the time-frequency technique, and it has been used in many practical applications like noise reduction and feature extraction. Wavelet time-entropy was used to extract features after wavelet packet decomposition and features that were extracted were used in the support vector machine for fault detection. Six different definitions about wavelet entropy and the wavelet entropy for identification of fault in a power system were used by He and Wang (2012). Wavelet entropy and statistical metrics for the features of ball bearings, and artificial neural network were used to diagnose faults of ball bearings (Bafroui and Ohadi, 2014). Žvokelj et al. (2010) presented the DT-CWT method for denoising the signal and its extraction for the diagnosis of weak gear fault. Lei et al. (2017) used the short time Fourier transform and then smoothed Wigner–Ville distribution for gear fault diagnosis.

20.2 OBJECTIVE

The objective of this work is to extract the meaningful information from the raw vibration signature by denoise and then extracting the hidden fault signature to distinguish between healthy and faulty gear states.

20.3 MAXIMAL OVERLAP DISCRETE WAVELET TRANSFORM (MODWT)

WT is capable of analyzing nonlinear and nonstationary signals. In WT, the signal is passed through a low-pass filter and a high-pass filter, which gives approximation and detail coefficients, respectively, and is downsampled by 2. The approximation coefficients are further decomposed into second-level detail and approximation coefficients, and so on up to the desired level of decomposition.

Discrete wavelet transform (DWT) has a revised version named the maximal overlap discrete wavelet transform (MODWT) (Cao and Xu, 2016), which overcomes downsampling process. There is no restriction on sample size in MODWT. The detail and smooth coefficients of MODWT are associated with zero-phase filters, wherein the temporal events and patterns in the original signal are consistent with the features in a multiresolution analysis (Yang et al., 2009; Shan and Li, 2010).

Denoising is the effective application of the WT, as it is capable of extracting the masked diagnostic information and intensifies the nonstationary and complex signal impulsive components.

Wavelet denoising mainly consists of the following steps:

- *Decomposition:* In wavelet denoising, the level of decomposition "N" is selected along with the selection of the mother wavelet.
- *Thresholding:* The next and most important step is to select the thresholding method. The wavelet coefficients are thresholded to eliminate the insignificant information from the signal.
- *Reconstruction:* At last the denoised signal is formed by reconstructing the coefficients by implementing inverse WT.

Wavelet denoising mainly uses two different philosophies of thresholding:

- *Hard thresholding:* In this philosophy, the coefficients that are below the threshold value are set to zero. In this type of thresholding, some useful information may be missed out as the coefficients that are below the threshold value are considered as noise and are discarded.
- *Soft thresholding:* In soft thresholding coefficients that are greater than the threshold value are subtracted from it. It is an extension of hard thresholding.

20.4 FEATURE EXTRACTION

Feature extraction is the process in which initial raw data can be reduced to more feasible groups for processing. Feature extraction of the signal identifies the dissimilarities between different faulty conditions. It is used as an input for automatic health monitoring and data-driven prognostics. During the failure of the machine or its elements, the energy and the distribution of the fault vibration signature change. These changes can be used for extracting features to identify different fault conditions efficiently.

20.5 EXPERIMENTATION AND METHODOLOGY

The experimental setup used in the present research work and the methodology adopted to extract useful information from machine vibration signals is presented in this section.

20.5.1 EXPERIMENTAL SETUP

For fault simulation in gears, a gearbox diagnostic simulator was used. Figure 20.1 shows the experimental setup used for the investigation. The test rig consists of a single, parallel shaft gearbox with rolling bearings and a programmable magnetic brake for loading. A 3 horsepower (HP) variable frequency AC drive with a variable frequency drive controller allows for adjusting the input shaft frequency. Root crack defects of magnitude 30%, 50%, and 70% were created by using wire electrical discharge machining (EDM). Figure 20.2 shows the healthy gear and gears with 30%

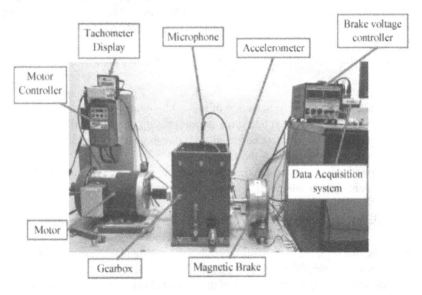

FIGURE 20.1   Test rig used for the experiment.

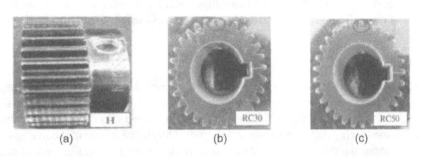

FIGURE 20.2   Gears with different conditions: (a) Healthy gear, H; (b) Gear with 30% fault; and (c) Gear with 50% fault.

and 50% root crack faults along with the nomenclature. The major parameters of the experimental test rig are given in Table 20.1.

TABLE 20.1
Test Rig Specifications

Parameter	Description
Rotational speed range	0–3000 rpm
External loading	Magnetic loading
Loading capacity	0.126–24.85 N-m
Gearbox speed reduction	Spur gear with single-stage speed reduction
Bearing	Deep groove ball bearing
Dimensions	Length = 100 cm, width = 50 cm, height = 60 cm

20.5.2 METHODOLOGY

Experiments were conducted at 13 Hz with 50% loading condition. The systematic methodology followed for the obtaining health condition indicator from the machine signature is given in Figure 20.3.

FIGURE 20.3 Experimental methodology.

20.5.2.1 Data Acquisition

The accelerometer, mounted on the bearing housing was used to capture the vibration signal for a running speed of 13 Hz at 50% loading condition. A data acquisition system (DAQ) was used for acquiring the raw data from the gearbox with a sampling frequency of 25,600 Hz.

20.5.2.2 Signal Denoising

The denoising algorithm was implemented on the wavelet toolbox of MathWorks MATLAB release 2020b. The level of decomposition N selected for this research is 4 as the theoretical gear mesh frequency, and its harmonics fall in the range of the selected decomposition level. So, four-level decomposition is executed and the "sym 4" wavelet is selected as the mother wavelet. The highest energy coefficient whose signal-to-noise ratio is maximum is selected for the analysis and then the denoised signal is obtained.

20.5.2.3 Feature Extraction

Eleven time domain statistical features, mean, standard deviation, skewness, kurtosis, peak to peak, root-mean square (RMS), crest factor, shape factor, impulse factor, margin factor, and energy, are utilized for fault diagnosis in this work.

20.5.2.4 Feature Selection

Of the eleven statistical features, the most prominent features are selected by using principal component analysis (PCA). PCA reduces the dimensions of the data sets, thereby preserving the maximum statistical information.

20.5.2.5 Health Indicator

The health indicator indicates the fault and distinguishes it from the severity of the fault.

20.6 RESULTS AND DISCUSSION

At first, the data were acquired when the machine was running in a healthy state for about 25 days for 1 second per day. After that incipient fault (i.e., RC30) starts occurring and data acquisition is done for another 25 days, and similarly for RC50 and RC70 for another 50 days, 25 days for each fault after that the faulty gear was replaced to avoid catastrophic failure.

The acquired data are then denoised using WT as discussed in the methodology section. Figures 20.4 and 20.5 depict the amplitude trend of raw vibration data and wavelet denoised data with fault progression, respectively. From these figures, it is noticed that in raw vibration data there is no significant indication of the increasing faults until the fault progression reaches RC70, whereas in denoised data, spikes are visible in the initial stage of fault progression (i.e., RC30 and RC50 are the acceleration amplitude, which is more prominent in case of RC70).

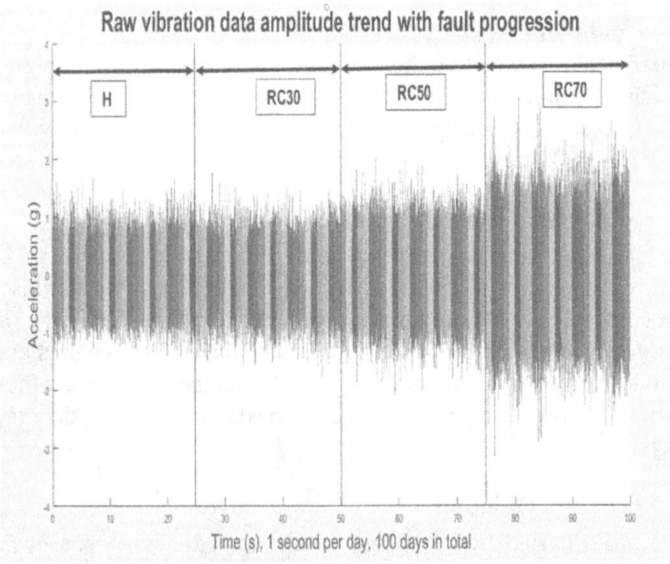

FIGURE 20.4 Vibration amplitude trend for raw vibration data.

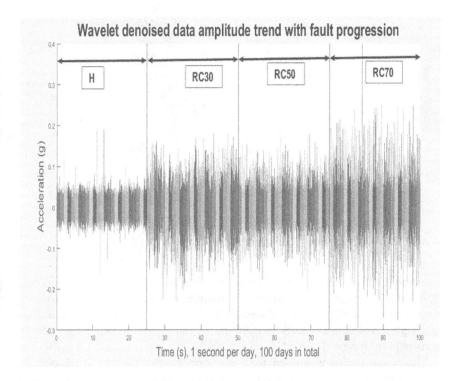

FIGURE 20.5   Vibration amplitude trend for denoised data.

Figure 20.6 is a zoomed view of the amplitude trend of denoised data for RC30. For more efficient differentiation between the faults, various features are extracted, as mentioned in Section 20.5.2.3. After feature extraction, the most prominent features are selected by feature selection using PCA. In PCA, raw and denoised feature sets are given as input and the most significant features obtained are PCA1 and PCA2 for both cases. PCA1 and PCA2 are trended for both raw

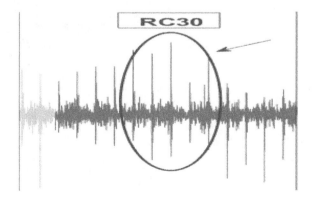

FIGURE 20.6   Zoomed view of amplitude trend of denoised data.

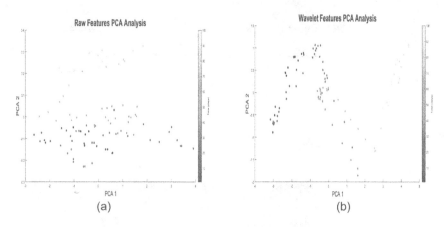

FIGURE 20.7   PCA analysis of (a) raw features and (b) wavelet denoised features.

and denoised signal as shown in Figure 20.7. The health indicator is the output of PCA, which consists of maximum information. For example, PCA1 carries the maximum information in case of wavelet denoised signature; thus, it is selected to be assigned as the health indicator. Similarly, raw signature PCA1 also contains maximum information.

The health indicator derived from raw and wavelet features is trended with time and they are shown in Figures 20.8 and 20.9. In Figure 20.8 for raw data, no

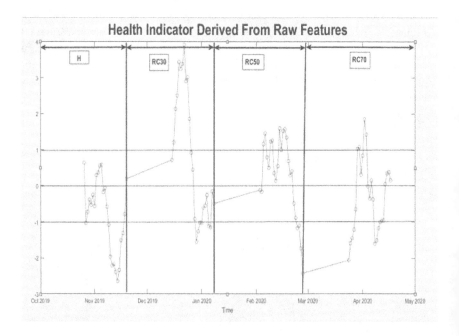

FIGURE 20.8   Health indicator derived from raw data.

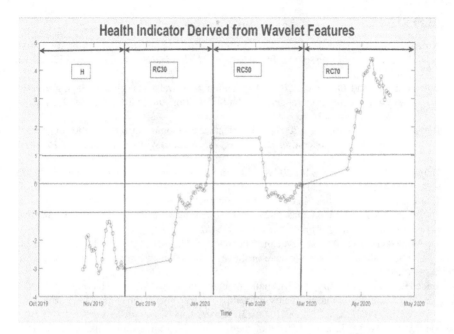

FIGURE 20.9   Health indicator derived from wavelet data.

significant trend is followed by the health indicator, whereas in Figure 20.9 for the denoised data the health indicator shows a monotone trend. For healthy condition, the health indicator value is near −1, and then it approaches 0 as faults are at a very initial level and the value of the health indicator is close to 1 for RC30. For RC50 the health indicator does not show an increasing response because RC30 and RC50 are not clearly distinguished. However, for RC70, there is a notable shoot-up of value, which indicates and segregates the fault very significantly.

20.7   CONCLUSIONS

This chapter presented an efficient technique for incipient fault detection in spur gears. Wavelet-based denoising of the vibration data of gears for different levels of root crack fault was done. In the proposed method, noise is removed from the raw signal by using MODWT in which symlet 4 is chosen as the mother wavelet and the level of decomposition selected is 4. After denoising, a total of 11 statistical features are extracted and the most prominent one is selected using PCA, which is taken as the health indicator for the present investigation. It was observed that with the denoising of the machine vibration signature the faults are more distinguished with the severity, which is not the case for raw vibration data as no trend is followed by the health indicator. Therefore, it may be stated that that denoising of the vibra-tion signals is essential for incipient fault detection in spur gears. The results of the analysis have demonstrated the cogency of the proposed technique for incipient fault detection.

REFERENCES

Antoni J., and Randall R.B., 2006. The spectral kurtosis: application to the vibratory surveillance and diagnostics of rotating machines, *Mechanical Systems and Signal Processing* 20(2), 308–331.

Bafroui H.H., and Ohadi A., 2014. Application of wavelet energy and Shannon entropy for feature extraction in gearbox fault detection under varying speed conditions, *Neurocomputing* 133, 437–445.

Bartelmus W., and Zimroz R., 2009a. A new feature for monitoring the condition of gearboxes in non-stationary operating conditions, *Mechanical Systems and Signal Processing* 23(5), 1528–1534.

Bartelmus W., and Zimroz R., 2009b. Vibration condition monitoring of planetary gearbox under varying external load, *Mechanical Systems and Signal Processing* 23(1), 246–257.

Cao G., and Xu W., 2016. Nonlinear structure analysis of carbon and energy markets with MFDCCA based on maximum overlap wavelet transforms. *Physica A: Statistical Mechanics and Its Applications* 444, 505–523.

De Azevedo H.D.M., Araújo A.M., and Bouchonneau N., 2016. A review of wind turbine bearing condition monitoring: state of the art and challenges, *Renewable and Sustainable Energy Reviews* 56, 368–379.

Endo H., and Randall R.B., 2007. Enhancement of autoregressive model based gear tooth fault detection technique by the use of minimum entropy deconvolution filter, *Mechanical Systems and Signal Processing* 21(2), 906–919.

He, Q., and Wang, J., 2012. Effects of multiscale noise tuning on stochastic resonance for weak signal detection, *Digital Signal Processing* 22(4), 614–621.

Heng A., Zhang S., Tan A.C., and Mathew J., 2009. Rotating machinery prognostics: state of the art, challenges and opportunities, *Mechanical Systems and Signal Processing* 23(3), 724–739.

Jia X., Zhao M., Di Y., Jin C., and Lee J., 2017. Investigation on the kurtosis filter and the derivation of convolutional sparse filter for impulsive signature enhancement, *Journal of Sound and Vibration* 386, 433–448.

Laha S.K., 2017. Enhancement of fault diagnosis of rolling element bearing using maximum kurtosis fast nonlocal means denoising, *Measurement* 100, 157–163.

Lee J., Wu F., Zhao W., Ghaffari M., Liao L., and Siegel D., 2014. Prognostics and health management design for rotary machinery systems—reviews, methodology and applications, *Mechanical Systems and Signal Processing* 42(1–2), 314–334.

Lei Y., Han D., Lin J., and He Z., 2013. Planetary gearbox fault diagnosis using an adaptive stochastic resonance method, *Mechanical Systems and Signal Processing* 38(1), 113–124.

Lei Y., Lin J., He Z., and Zuo M.J., 2013. A review on empirical mode decomposition in fault diagnosis of rotating machinery, *Mechanical Systems and Signal Processing* 35(1–2), 108–126.

Lei Y., Liu Z., Ouazri J., and Lin J., 2017. A fault diagnosis method of rolling element bearings based on CEEMDAN, *Proceedings of the Institution of Mechanical Engineers, Part C: Journal of Mechanical Engineering Science* 231(10), 1804–1815.

Peng Z.K., and Chu F.L., 2004. Application of the wavelet transform in machine condition monitoring and fault diagnostics: a review with bibliography, *Mechanical Systems and Signal Processing* 18(2), 199–221.

Randall R.B., 2011. *Vibration-Based Condition Monitoring: Industrial, Aerospace and Automotive Applications.* Wiley, Hoboken, NJ.

Samuel P.D., and Pines D.J., 2005. A review of vibration-based techniques for helicopter transmission diagnostics, *Journal of Sound and Vibration* 282(1–2), 475–508.

Samuel P.D., and Pines D.J., 2009. Constrained adaptive lifting and the CAL4 metric for helicopter transmission diagnostics, *Journal of Sound and Vibration* 319(1–2), 698–718.

Sawalhi N., Randall R.B., and Endo H., 2007. The enhancement of fault detection and diagnosis in rolling element bearings using minimum entropy deconvolution combined with spectral kurtosis, *Mechanical Systems and Signal Processing* 21(6), 2616–2633.

Shan P.W., and Li M., 2010. Nonlinear time-varying spectral analysis: HHT and MODWPT, *Mathematical Problems in Engineering*, 2010, 618231.

Shao Y., and Nezu K., 2005. Design of mixture de-noising for detecting faulty bearing signals, *Journal of Sound and Vibration* 282(3–5), 899–917.

Tan J., Chen X., Wang J., Chen H., Cao H., Zi Y., and He Z., 2009. Study of frequency-shifted and re-scaling stochastic resonance and its application to fault diagnosis, *Mechanical Systems and Signal Processing* 23(3), 811–822.

Vanraj, Dhami S.S., and Pabla B.S., 2017a. Non-contact incipient fault diagnosis method of fixed-axis gearbox based on CEEMDAN, *Royal Society Open Science* 4(8), 170616.

Vanraj, Dhami S.S., and Pabla B.S., 2017b. Optimization of sound sensor placement for condition monitoring of fixed-axis gearbox, *Cogent Engineering* 4(1), 1345673.

Vanraj, Singh R., Dhami S.S., and Pabla B.S., 2018a. Development of low-cost non-contact structural health monitoring system for rotating machinery, *Royal Society Open Science* 5(6), 172430.

Vanraj, Dhami S.S., and Pabla B.S., 2018b. Hybrid data fusion approach for fault diagnosis of fixed-axis gearbox, *Structural Health Monitoring* 17(4), 936–945.

Wang L., Ye W., Shao Y., and Xiao H., 2019. A new adaptive evolutionary digital filter based on alternately evolutionary rules for fault detection of gear tooth spalling, *Mechanical Systems and Signal Processing* 118, 645–657.

Yan R., Gao R.X., and Chen, X., 2014. Wavelets for fault diagnosis of rotary machines: a review with applications, *Signal Processing* 96, 1–15.

Yang Y., He Y., Cheng J., and Yu D., 2009. A gear fault diagnosis using Hilbert spectrum based on MODWPT and a comparison with EMD approach, *Measurement* 42(4), 542–551.

Zimroz R., and Bartkowiak A., 2013. Two simple multivariate procedures for monitoring planetary gearboxes in non-stationary operating conditions, *Mechanical Systems and Signal Processing* 38(1), 237–247.

Žvokelj M., Zupan S., and Prebil I., 2010. Multivariate and multiscale monitoring of large-size low-speed bearings using ensemble empirical mode decomposition method combined with principal component analysis, *Mechanical Systems and Signal Processing* 24(4), 1049–1067.

Index